ARTIFICIA

SHAPE THE SIX PILLARS OF OUR LIVES

Babak A. Taheri, Ph.D.

For Maryam and Michelle

With Special Thanks to Sam Lim for his support

About the Author

Dr. Babak Taheri

President and CEO of IBTechnologies.

Dr. Taheri is the president and CEO at IBTechnologies, a leading Merger and Acquisition partner with venture firms, private equity firms, and hedge fund managers. He also volunteers as a board member of the Parisi House on the Hill and chairs the advisory board of the electrical and computer engineering department at U.C. Davis.

Prior to IBTechnoloiges held several positions as CEO/CTO and Member of the Board of Directors at Silvaco Group, VP & GM of the sensor solutions division at Freescale Semiconductor (now NXP), VP &GM at Cypress Semiconductor (now Infineon), and VP of Engineering at Invensense (now TDK).

Babak was the recipient of "the perfect project award" in 2003 while at Cypress; Twice recipient of the "Diamond Chip Award" in 2013 /14 while at Freescale; recipient of the MEMS & Sensors executive of the year award in 2014, and in 2015 was the recipient of the Distinguished Engineering Alumni Medal from UC. Davis College of Engineering.

He also held key roles at SRI International and Apple. He received his Ph.D. in biomedical engineering from UC Davis with majors in EECS and Neurosciences, has over 20 published articles, and holds 29 issued patents.

Preface/ Prologue

This book is the accumulation of many years of research into practices and approaches used in the new connected world. Specifically, it focuses on the role that sensors play in our lives by changing business models and how the valuable data created, are being utilized by us and monetized by the businesses.

Did you know that since the early 80s, we have adopted new advanced technology and products about 100x faster than the industrial revolution? There are three major reasons why:

First, the exponential population growth to 9 billion by 2025 will keep driving demand for goods. Second, products are becoming intelligent and connected to the Internet. Product-generated data, which is in massive amounts, is collected on databases, and the insights generated by these are then analyzed, shared, and socialized. The number of Internet-connected devices is forecasted to be 75 billion by the year 2022 worldwide. This is about eight times the human population in the same year (8 devices/person on average). By 2025, this number is predicted to exceed 500 billion units of connected devices, which will be about 55 times larger than the world's population (9.1 billion), representing another exponential growth.

Third, in the past decade, the usage of sensors in consumer electronics (e.g., mobile phones, tablets, laptops, cameras, security systems), large industrial plants (e.g., factories, agriculture, and hospitals), automobiles (e.g., passenger cars, trucks transitioning to autonomous vehicles), and medical devices (e.g., wearables, mobile devices), have increased significantly. As a consequence of using sensors, besides the substantial impact on all aspects of our lives, the world population is generating more data than ever before. We will exceed 177 Zettabytes ($1ZB = 10^{21}$ bytes) of data by 2025. A Zettabyte is equal to the amount of data that can be stored on 250 billion DVDs. In turn, this rate of data generation is increasing demands for computation, storage, software, analytics, and sensing even further.

Sensors have enabled us to collect meaningful data from our environment, our goods, and ourselves. This data is converted to information to help us make better-informed, data-driven decisions. The sensors' data is also combined (sensor data fusion) with other data sources to generate the necessary information required for the analytics and context awareness.

For example, in a passenger car equipped with Advanced Driver-Assistance System (ADAS), the camera sensors and/or radar sensors collect the environment image data. It is then combined with other information such as speed, tire conditions, road conditions, the distance of objects, and the breaking-ability of a car to avoid a collision.

Due to the complicated nature of their design and manufacture, sensors have been getting traction in high-volume products only in the last 10 years. Currently, sensors like Integrated Circuits (ICs) and printable electronics are being mass-produced at very economical costs. Did you know that the average smartphone in 2017 was equipped with over ten sensors (with less than $5 bill in material cost), and the average passenger car was equipped with over fifty sensors (with less than $200 bill in material cost)?

On January 9th, 2007, when Apple introduced the first iPhone, there were less than 4 sensor types included: touch sensors for display, an image sensor for the camera, a microphone for voice, and an accelerometer for detecting the phone's linear motion. The iPhone X, introduced in 2017, has over 14 types of sensors, which include 3D image sensors and fingerprint sensors. Essentially making the phone a sensor hub.

Sensors make products more human by mimicking the human sensory functions, making the products more functional, more appealing, helping us to understand our environment and ourselves. They range from the image sensors in cameras (eyes) to microphones (ears) to speakers (voice) – enabling the products to see, hear, and sense the environment (motion and chemical sensors) around them. This massive data generation, in turn, also enables many new products and services.

The products/technologies and services that appeal to our humanity tend to have social currency, transform our lives, and impact the way things are done that were previously unimaginable. This can best be described by Steve Jobs, "Every good product I've ever seen is because a group of people cared deeply about making something wonderful that they and their friends wanted. They wanted to use it themselves." Sensors help products do this.

This book focuses on the impact sensors have had on products while providing a perspective on how integrating them with other technologies can transform industries, companies, and our lives.

Let the journey begin!

Contents

Page Left Blank Intentionally

Introduction

"They also explained how the sensors can monitor the levels of Acetone on people's breath, and this can be used to tell people who suffer from diabetes when their next insulin shot is due."

Anne C. Campbell (English Politician)

"As sensors and networks continue to expand around the world, we'll see violence drop even further. After all, when there's a danger that your actions can be caught on tape and shown around the world, you're more responsible for your behavior. "

Peter Diamandis (Engineer, Physician)

By 2013, in a span of five years, I personally had gone through four tablets, three upgraded internet routers, five mobile phones, three laptops, two desktops, two-set of home security systems, two cars, and had upgraded my computers to six Terabytes (TB) of storage plus additional storage on hosted cloud. I had stored two decades' worth of pictures, videos, digital books, music, and movies.

Compared to my personal storage use, as of 2013, the sum total of data held by all the big online storage and service companies like Google, Amazon, Microsoft, and Facebook is estimated to be 1.2 Exabyte (EB $=10^{18}$), grown 10 Zettabytes (ZB $=10^{21}$) by 2020.

Further, my mobile devices (phones, tablets, watches, wrist bands, and laptops) over the span of ten years had incorporated over twenty sensors and my car over a hundred sensors.

An iPhone 7 has 12-megapixel and 7-megapixel image sensors for its cameras, a fingerprint sensor for access control, a barometer that can measure pressure in the environment as well as elevation. A three-axis gyroscope that can measure rotational motion sensing with potential use in optical image stabilization (OIS) of the camera(s), GPS-assist, and gaming control. A three-axis accelerometer for linear motion sensing and gaming function. A proximity sensor for knowing when the phone is near your ear to turn off the screen and save battery power and avoiding the accidental selection of functions, and an ambient light sensor for screen-dimming to save power when the ambient light is high, as well as a touch sensor for the screen, microphones, and speakers.

Another example is, the iPhone 12 Pro that has over 10 sensors listed below:

- MagSafe 15-watt wireless charging that senses charger presense
- 12 MP triple camera system with wide-angle $f/1.6$ (OIS) image sensor
- 12-MP ultra-wide angle $f/2.4$, image sensor
- 12-MP telephoto $f/2.2$ (OIS) cameras
- LiDAR module using time of flight sensing to measure distance/depth
- The loudspeaker
- Taptic Engine for vibration notifications
- Accelerometers – for orientation and navigation (nav)
- Gyroscopes for orientation, image stabilization, and potential nav
- Ambient Light sensor for display intensity control
- Proximity sensors for screen/touch on/off during phone calls
- Barometer for pressure measurements
- Digital Compass (historically a magnetometer or magnetic sensor)

My modern car has four oxygen sensors, several fuel pressure sensors, Intake Air temperature and airflow rate, throttle position, ambient air temperature, tire pressure, passenger occupancy, and tens of other sensors for tens of other purposes.

All these products are equipped with sensors, constantly monitoring and capturing data that is stored, partially or fully analyzed, and connected to the Internet (Internet of Things) for storage, further analysis, sharing, or decision-making.

We have just begun this new journey, and the waves of change are coming. This is the next big thing from the sensor usage, and the data generated by these sensors will expand markets in storage, computation, data analytics/information processing, and artificial intelligence—to extract information within a context to help us make data-driven and informed decisions.

This new sensor-connected world will empower individuals even more by providing real-time information with real-time feedback, so we have more say in our daily lives. We will know how energy is consumed, what safe and high-quality foods are available/consumed, how we have more say in our health management, and how local/remote environments influence us.

For example, if at any given time, we know how much energy each appliance in the house is using, we can decide which light fixtures to dim or turn on/off and which appliance to turn on/off, or put in a lower power setting, whether we are at home or not. Similarly, we will have access to our bodies ' vital signs in real-time on our mobile devices. It will enable us to make decisions on medicine intake and monitor the effect of medicine, food, and beverages we consume. We moved the data generation from laptops/desktops a decade ago to our sensor-enabled wearable, mobile devices, homes, cars, and buildings. This start point of data generation is what I call the "Sensor Edge Node" (SEN).

Sensor edge nodes with intelligent computational nodes that have storage, connectivity, and software are shaping the new information technology paradigms. They are enabling mobility and, more importantly, leveling the playing field across the world. The world will become a more level-playing field for individuals and entities, thanks to equal access to information, analysis, control, connectivity, and decision-making tools.

This book is meant to be a guide to understanding and interpreting new exponential technologies such as sensors and their impact when combined with computing, storage, connectivity, and related software. It is written to provide adequate background on both technology and business aspects. This book will explain the very complex nature of technologies, with the aim being the simple, thoughtful awareness of readers with varying technology and business backgrounds.

The questions that this book addresses include but are not limited to: What sensor technologies are worth adopting that help products appeal to our humanity? How do these new technologies transform our lives by addressing the needs of at least one or all the six pillars that define our humanity? Those six pillars being:

1) Our Health- Mind and Body

2) Our Food - Safety/Quality from farms to homes

3) Our Shelter(s) – Safety/Energy Efficient

4) Our Mobility – Secure Access/Control from anywhere, any time

5) Our Energy Resources – Oil, electricity, Hydrogen

6) Our Freedom – Safety/Security/efficient use of time

This book also gives specific examples and future trends in each chapter.

The ideal product/technology and service that addresses all the pillars of our lives are the product of choice. In reality, not all do and still succeed in it; this will be covered in each chapter as well.

Chapter 1 will give you a background on the pyramid of value chains. The more pillars that products address, the higher levels of pyramid value chains, the more differentiated and useful they become. This value chain is the blueprint for any of the technologies that are mentioned in this book and specifically addresses what sensors are and how sensors impact the value chain when they become intelligent, connected, providing real-time information for decision-making.

Chapter 2 will cover how sensors can help us monitor, analyze, provide information to patients, healthcare providers, and ourselves. This, in turn, is shifting the healthcare model to a more patient-centric model. It will cover remote health monitoring, remote interaction with physicians, to social medicine applications for mobile devices. Examples of products are given with potential improvements, moving products up the pyramid of value chains. It covers what I call from "womb to tomb" sensing and analytics.

Chapter 3 will cover smart environments pertaining to food safety, food quality, food production, food transportation, and food storage. What are these environments like, which ones are being worked on, and which ones need more attention to provide us with lower cost and safer and higher quality food? It covers what I call "harvest to home" sensing and analytics.

Chapter 4 will cover how sensing our homes, buildings, and infrastructure play a role in many of our daily decision-making. It covers smart buildings and what I call "secure and aware buildings" sensing and analytics.

Chapter 5 will cover mobility, including data accessibility within our mobile devices that not only include phones, wearables but also include intelligent, self-driving vehicles. Having access to our data and information around the clock is what I call "anywhere and anytime" kind of data sensing.

Chapter 6 will cover the energy sources available and how sensors help optimize energy usage while we transition to new energy sources such as solar, wind, hydrogen, and hybrid energy models. I call this "predictable energy" sensing.

Chapter 7 will cover how access to all the sensor data and information enables us to make better decisions securely and free us from unproductive tasks that burden our freedom. Sensors also enable us to detect security threats and give us real-time information to make the world a safer place. I call this "Predictable security and safety" through sensing.

Chapter 8 will summarize and provide a glimpse of where we are going with the technologies that impact our lives with special sections on impact of Covid-19.

Chapter 1: Pyramid of Value Chains and Sensors

"We tend to overestimate the effect of a technology in the short run and underestimate the effect in the long run."

Roy Charles Amara (1925 - 2007)

"War is ninety percent information."

Napoleon Bonaparte (1769 –1821)

1.1 How value increases by monitoring, automation, and real-time control

The technological advances in the top market segments of Financials, Digital Technology, Health Care, Consumer Goods, Electronics, and Energy, are the driving forces for changes as we see them today and in the unforeseen future. Although the advances in each market vary, they all share the pyramid of value chains in order to deliver more functionality, lower cost, smaller size, individualized, and ubiquitous value to the end-user.

1.2 The Pyramid of the value chain for the connected world

The need to collect data for the successful execution of tasks (e. g., investment, and market expansions) has been understood and valued for several centuries. *The goal has been to turn data into information and information into knowledge to help in the decision-making process.*

The evolution of the rate of data collection, type of data, and how it is collected and analyzed has evolved and is rapidly changing. We have come a long way from manual data collection of limited data types. It is now all about mega data: the unlimited data types and their automatic processing, sharing, and analysis.

7

Just a decade ago, most traders relied on published newspaper data to perform long and short-term investments. Today, individuals can place trades in real-time 24/7 on any of the world's markets with little or no intervention. Mobile apps can take the trade, place limit thresholds on gain or loss, and as soon as those thresholds are met, the trade is placed.

An extreme example of this is the high-speed trading that takes place daily by intelligent computers collecting and analyzing market data and placing trades at the speed of light. High-Frequency Trading (HFT) is the way of present and future trading and accounts for 50% of U.S trading. According to Capgemini, *"in 2010, HFT was estimated to have accounted for 56% by volume of the entire equity turnover in the U.S., up from 21% in 2005."*

Exhibit 1, below from TABB Group, shows HFT as a percentage of all U.S. Equity Trading from 2005 through 2016.

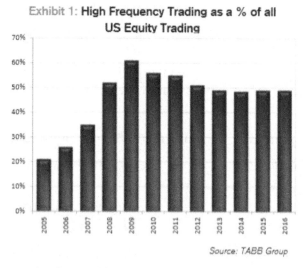

Exhibit 1: **High Frequency Trading as a % of all US Equity Trading**

Source: TABB Group

Like any new technology, although HFT has numerous advantages, it did create the infamous "May 6th Flash Crash" in 2010, resulting in regulatory reactions and corrections. That incident indicated that better regulation and more intelligence on the control is necessary, as well as real-time correction of data, for such technologies to be effective in the long run.

Historical evolution of how sensor data is collected, stored, analyzed, shared can also be explained by the four levels in the pyramid of value chain shown in figure 1.2.

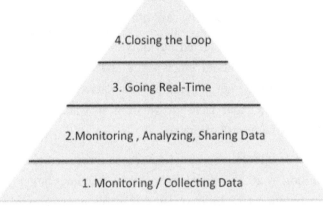

Figure 1.2: Four levels in the pyramid of value chains

The higher levels of the pyramid initially command fewer participants since they require more technical and marketing enablement, hence a slower adoption rate.

At the lower levels, more participants adopt the technologies offered. They are simpler and mimic the behavior of what the users are familiar with, such as storing data. But as the value at the higher level of the pyramid is realized, the same or fractions of the lower-level participants will gradually adopt the newly created values and move to higher levels of the pyramid.

Breakdown

Level 1 of the pyramid represents the monitoring/collection and storage of data locally, on clouds, and in data centers. Currently, many large companies, such as Amazon, IBM, Microsoft, Apple, Google, and Facebook, provide cloud services that make up a large portion of their revenue. These and many smaller companies provide data storage on the cloud for individuals to store wearable data, pictures, videos, and other data as backups or added storage for their devices. At level 1, data is viewed as "property," and storing that data is taking the analogy of property storage units that one would rent monthly for keeping extra stuff they need but do not have the storage space at home for. We access our storage unit property less often than what we have at home unless the access is easier (as in the case of the cloud).

In level 1, the data is collected by individuals or devices and *stored* in devices, and the cloud without much value added to convert the data into information. Due to the high demand for raw data storage that is accessible from anywhere, participants have grown from all walks of life to use such services and are continuously growing.

The number of Internet-connected devices is forecasted to be in the range of 25 to 75 billion by 2020, which will be 3.24 to 9.7 times larger than the then worldwide population (7.7 billion). By 2025, this number may exceed 500 billion units of connected devices, 55 times larger than the then worldwide population of 9.1 billion. The result of this growth is the exponential expansion of data to 6.44 Zettabytes ($1ZB = 10^{21}$ bytes).

As a reference, a Zettabyte is equal to the amount of data that can be stored on 250 billion DVDs. We are well short of this number for the current storage capacity in the world. Level 1 participants take advantage of available storage that is partially free of charge. As their need grows for a higher capacity of storage, they pay the fee for the additional storage space. A large portion of this stored data comes from sensors. Image sensors in phone cameras, digital cameras, and security systems contribute to the majority of data stored in public and some private clouds.

Level 2 of the pyramid represents the monitoring, *analyzing*, **and sharing** of the data for creating new value. Currently, as individuals, we have very little access to sophisticated analytics software for our stored data. However, we know larger companies such as GE, IBM, and Google, provide services to store and analyze data, including cognitive, psychological, and behavioral analyses of markets and users. But now, this is getting more personalized at the individual level.

Although one might store thousands of photos on the cloud, finding the right photo when you need it is a daunting challenge, given the thousands you have to sort through. No service or program is provided to organize, search, and enable more useful information about the photos, such as types of photos, their location, and content. By creating software that analyzes the photos and provides more information to us, we can benefit from a better experience beyond just storing the photos. We can actually find what we need when we need it. Similar to what Google does as a search engine, the search for graphs can have its own intelligent sub-search engine.

For example, if the software can look at each picture and detect where the photos were taken, check its time-frame, content, and allow the user to sort, search, organize, and share the photos as they desire can make the experience more valuable. This software that does the photo analysis and detection is at level 2 of the pyramid. As it turns out, Google is working on such software and has announced the release of its beta version. The way Google describes this capability that is added to its cloud storage service is as follows.

"Google Cloud Vision API enables developers to understand the content of an image by encapsulating powerful machine learning models in an easy-to-use REST API. It quickly classifies images into thousands of categories (e.g., "sailboat," "lion," "Eiffel Tower"), detects individual objects and faces within images, and finds and reads printed words contained within images. You can build metadata on your image catalog, moderate offensive content, or enable new marketing scenarios through image sentiment analysis."

Other companies are doing similar work on data sets such as GPS location data, web browsing habits, health activities, shopping habits, and many more that users are willing to pay for the data analytics aspects of their services.

Another example of level 2 participation is storing and analysis of location data on fleets for asset tracking. Analyzing that data to provide information on fleets' efficiency, scheduling, cost-reduction, and energy consumption is an example of how knowledge is increased about the data set, and hence more value and new services are created for monetization. A combination of GPS, compass sensors, and motion sensors enables such activities.

The Big Data technology and services market is a worldwide fast-growing multibillion-dollar business. A recent IDC forecast shows that the Big Data technology and services market will grow at a 26.4% compound annual growth rate (CAGR) to $41.5 billion through 2018, or about six times the growth rate of the overall information technology market.

By 2020, IDC believes that line of business buyers will help drive analytics beyond its historical sweet spot of relational performance management to the double-digit growth rates of real-time intelligence and exploration/discovery of the unstructured worlds. In 2015, the top companies will be generating hundreds of million dollars in data analytics through a combination of sales of software applications, consulting services, Information Technology (IT) services, data warehousing, hardware for computation-storage-connectivity included companies such as PWC, part of the Price Waterhouse Coopers business, Accenture, Palantir, SAS Institute, Oracle Corp, Teradata (2007 sup off from NCR), SAP SE, DELL, Hewlett-Packard Co, and IBM.

For example, although IBM's overall revenues have been flat or declined for ten consecutive quarters (as of November 2014), the company has momentum in the Big Data market -- generating $1.37 billion in revenues for 2013, as Wikibon estimates. Those revenues were split across hardware (31%), software (27%), and services (42%). Eager to potentially accelerate those Big Data revenues, IBM in late 2014 said it was expanding cloud-focused releases of Cognos while also working more closely with Watson-focused software. The creation of analyzed data and sharing of the information with a large number of individuals and groups is where the focus has become.

IBM and Apple have announced a collaboration on data analytics for iOS and Apple's mobile platform to bring the sharing of data to individuals and companies that use iPhone, iPad, and wearable devices. As Ginni Rometty, IBM Chairman, President, and CEO, said in the joint press release of July of 2014,

"Mobility—combined with the phenomena of data and cloud—is transforming business and our industry in historic ways, allowing people to re-imagine work, industries, and professions. This alliance with Apple will build on our momentum in bringing these innovations to our clients globally and leverages IBM's leadership in analytics, cloud, software, and services. We are delighted to be teaming with Apple, whose innovations have transformed our lives in ways we take for granted but can't imagine living without. Our alliance will bring the same kind of transformation to the way people work, industries operate, and companies perform."

Level 3 of the pyramid is the level at which the data monitoring, collection, and analysis are done in real-time in fractions of seconds. Only a few percentages of the companies listed previously will participate in this area due to the complexity, technology, know-how, and cost of infrastructure needed to realize it.

At this stage, data analytics is provided in an open-loop fashion to the end-user, with small and no real-time feedback from the user(s) to the data source. This is also known as dynamic analysis and static feedback. The entities that have embraced this level of value service include the military, the emergency services, the financial companies, and trading companies since real-time data makes a significant difference in the outcome of their decisions.

The next level of the value chain is level 4 of the pyramid that includes closing the loop between the analytics and the feedback from the user and is ideally real-time. This is also referred to as dynamic analysis and dynamic feedback. For example, in emergency responses to disasters, the data from the disaster site, while it occurs in real-time, is provided to the cloud by people and sensors. The analytics provide guidance on mobile devices to responders on the disaster site, and as the responders enter the feedbacks, the next set of analyses is done, and this continues until the situation comes under control.

The highest level of the pyramid includes Real-time Control. For example, in military applications where the sensors provide site information and guide planes and drones to deliver supply and ordinance to the targeted area(s).

In this case, the actuators in control of the delivery system are under the supervision of real-time sensor data from the site. A commercial usage example for this level of value is being considered now for drone delivery of mail and other types of packages via such a system. While the position, altitude, and longitude of the drone from sensors are sent to a centralized location, guiding the drone is done partially remotely and partially automatically. Amazon and Alphabet (a.k.a. Google) have been investing heavily in these types of systems since the value of service they can provide would be at the lowest cost point compared to traditional delivery of goods. In August of 2016, **Domino's** demonstrated that it could deliver pizza by drones in New Zealand. In December of 2016, **Pizza Hut** in Shanghai, China, demonstrated a Service Robot working in the stores and servicing customers instead of a person.

Another example is the significant investment made by the automotive and transportation companies for autonomous cars, trucks, trains, ships, and airplanes to enable the movement of goods, people, and services that require real-time analytics and control. In March of 2016, **Domino's** demonstrated Pizza delivery by an unmanned vehicle. In May of 2017, **Volvo** tested automatic garbage collection by self-driving trucks. In 2016, a patent awarded to **Google** suggested that they are planning package delivery by self-driving trucks.

The human population has grown exponentially from the 1800s (about 1 billion) to the present time (7 billion). We are due for another industrial and social revolution. We cannot afford to do things the way we have for 200 years and the ways we still are, using the limited resources available to us. Both natural resources and skilled human resources are limited. The necessity for improving resource utilization has become more relevant with population growth. We need to move into a proactive way of doing things than reactive to optimize resource utilization. The way we care for our bodies, the environment (pollution, global warming), the ways we grow/raise our food (genetically modified vs. organic), the ways we consume energy (fossil fuels vs. Hydrogen), water/other resources, and the ways we travel /commute, all of them need to be reinvented. In order to make the necessary changes in how we do things, we need to have more real-time information about them to be able to make more informed decisions about them.

For us to obtain real-time information about things, events, and resources, we need to collect data on them in real-time and decipher the data as we go to make sense of it; doing so will help us make an informed decision. To get data from things, events, the environment, and resources, we need to have sensors that transfer the data from things, events, environments, and resources to computers that are used for storage and analysis. Most of this will be done through the Internet, cloud storage, and cloud or distributed computing.

1.3 Biological and Artificial Sensors

In the broadest definition, a sensor detects events or changes in its environment (energy) and then converts it to another form of energy for processing. To humans and animals, perception of the world is through our biological sensors. For humans and most animals, vision, auditory (hearing), somatic sensation (touch), gustatory (taste), olfaction (smell), and vestibular (balance/movement) senses are the main sources of observation and learning. For example, the eye converts photons (light Energy) to electrochemical pulses (another form of energy) for the brain to process and analyze.

1.3.1. Biological Sensors

More advanced biological sensors can be found in some animals with differing and/or superior performance to that of humans; they evolved in different environments suitable for survival, some of which are quite interesting. Did you know that the Silvertip Grizzly Bears are able to smell things from 20 miles away and across time? Similar to that of a bloodhound dog that knows who walked down the street from several days ago. As long as the scent remains in the air, they can detect it and even tell when the scent was placed in that location.

The human eye can see three main colors of red, green, blue (three-color cone cells in the retina) and has about 167 degrees horizontal and 150 degrees vertical fields of view. The Jumping Spiders have 360 degrees view of their surroundings due to the placement of their four eyes and are capable of seeing four colors (tetrachromatic vision – Four color cone cells in their retina) that include ultraviolet too. This enables them to hunt insects even in the dark. Humans can use instruments to extend their light-sensing spectrum by use of a UV photodiode in special cameras, which can detect the 240-370nm range of light (which covers UVB and most of the UVA spectrum). If you have not tried UV goggles, I recommend them. Shown in the picture below, on the left-hand side, is the view of what a human sees when looking at a Monarch Butterfly, while the right-hand image shows the view of the same from an animal having an ultraviolet vision.

Figure 1.3: The view of what a human sees when looking at a Monarch Butterfly

Bees find their hive with the help of a small ring of magnetite particles, magnetic granules of iron, inside their stomach. Bees can detect the magnetic field of the Earth that defines its location and hence find their destination. By the way, the compass in your mobile phone consists of a magnetic sensor that detects the magnetic field of the earth similar to that of the bee's sensor.

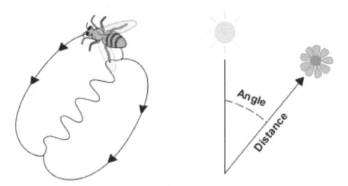

Figure 1.4: How bees detect magnetic sensor

Biological sensors have always been the aspiration for designing artificial sensors. Similar to biological sensors, without artificial sensors, our perception of the world would be limited in the connected world. The companies that understand the Internet of Things (IoT) and strategic relevance of sensors when combined with computing, communications, and analytics are winning in the marketplace (e.g., Amazon, Google, Apple, Tesla).

1.3.2. Artificial Sensors

Today, artificial sensors are used in everyday objects, and their applications are growing. We use touch-sensitive buttons and displays to interact with phones, displays, lamps, phones, thermostats, refrigerators, and the infotainment in our cars. Through a communication medium and IoT software, intelligent sensors alert you when the taps in your bathroom leak or when there is smoke in your living room. They also light up, set the thermostat, open/close the garage door, or remotely monitor the house.

To make any home a smart home and to make any device a smart one, sensors have become an integral part of the electronic products by attaching them to a microprocessor and, ideally, a wireless interface. The companies that succeed and realize such wireless sensor hubs will have a higher demand for their products (e.g., Amazon, Google, Apple, Tesla).

The very basic Internet of things (IoT) is to have, at the minimum, an *intelligent connected* device.

Collecting sensor data and transferring the sensor's data to a local processor and cloud computing enables faster and much in-depth analytics that are needed for a large amount of data from various geographical locations. The data is collected locally from each sensor but globally looked at while both information and decision criteria are provided.

The artificial sensors are usually classified based on the energy/parameter(s) that they sense. The top 10 sensors that are highly utilized and are forecasted to do so in the next five years are shown in the figure 1.5.

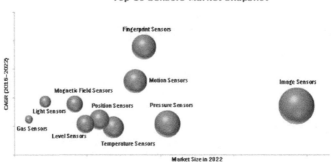

Source: Industry Experts, Secondary Research, and MarketsandMarkets Analysis

Figure 1.5: Sensor Market Snapshot

The vertical axis in the figure represents relative compounded annual growth (CAGR) for the span of 2016 to 2022, the horizontal axis shows the relative market size in 2022, and the bubble size shows the relative volume. The ranking based on the volumes is 1. Image Sensors, 2. Pressure Sensors, 3. Fingerprint Sensors, 4. Motion Sensors, 5. Temperature Sensors, 6. Position sensors, 7. Level Sensors, 8. Magnetic Field Sensor, 9. Light sensors, and 10. Gas Sensors.

A high-level use case, market size, and technologies for the top three sensor types (Image Sensors, Pressure Sensors, and Fingerprint sensors) are given below. However, examples of each type of artificial sensor are given in each chapter based on their use case, and their markets are discussed accordingly.

Image Sensors (IS): There are several types of image sensors that either produce red/green/blue (RGB) images, black and white images, thermal/heat images, infrared images, and/or X-ray images. Do you recall taking an X-ray in the '90s, and it would take hours to process the X-ray film? That is not the case anymore. When X-ray image sensors are used, X-ray images are available immediately to you and your doctors.

Although there are various technologies that have evolved over the years to realize image sensors, the majority of the markets (mobile phones, cameras) are dominated by Complementary Metal Oxide Semiconductor (CMOS) technology. These sensors are the eyes that take pictures, movies and are installed in cell phone cameras (1-3 per phone), digital cameras in building perimeters for security, on Drones/un-manned aerial vehicles (UAV) for surveying, in satellites for survey/security, robots for machine vision, toys for user-experience, automobiles for Advanced Driver Assistance (ADAS), and in tablets (figure 1.6).

Fig 1.6: Security, Photography, Mobile Phone, and ADAS camera

These image sensors vary in the number of pixels and size (shown below)—from few pixels to millions of pixels, enabling you to expand and print images that are wall-sized or provide small XGA to 1080p, 4K, or 8K high-definition movies.

Figure 1.7: 8K, 4K AND XGA resolution inage sensors

Also, the camera module size varies from pinhole camera for toys/security monitoring to basketball-size used in satellites and telescopes that take photos and movies from outer space. Moreover, their speed of operation varies from a few frames per second (fps) to hundreds of fps allowing high-speed photography.

Each pixel of an image sensor converts photons to electrical signals that are then processed by a microprocessor, converting the photon information to still images or videos. Similar to the biological sensors and depending on the application, a pixel is designed to convert light of various wavelengths (color/black and white, x-ray, infrared, thermal, ultraviolet...) to electrical signals. The top companies that make image sensors for the visible light spectrum are shown in the figure 1.8. Both their 2014 and 2015 annual revenues are shown as well. In 2016 and 2017, an additional 10% growth for this market was achieved mainly due to applications in robotics, drones, UAVs, toys, and security.

Shown below (Yole development) are the ranking of major image sensor players and their revenue in 2014 and 2015, showing growth from $9.3 B in 2014 to $10.3 B in 2015.

CMOS image sensor player rankings 2016

	Company	2014 (in $M)	2015 (in $M)	YoY Growth (%)
1	Sony	2,779	3,645	31
2	Samsung	1,825	1,930	6
3	Omnivision	1,378	1,250	-9
4	On Semicondutor	670	810	21
5	Canon	482	404	-16
6	Toshiba	360	350	-3
7	Panasonic	244	336	38
8	SK Hynix	200	325	63
9	Galaxycore	325	275	-15
10	STMicroelectronics	260	200	-23
11	Pixart	166	170	2
12	Pixelplus	114	130	14
13	Other	498	523	5
		9,300	10,348	

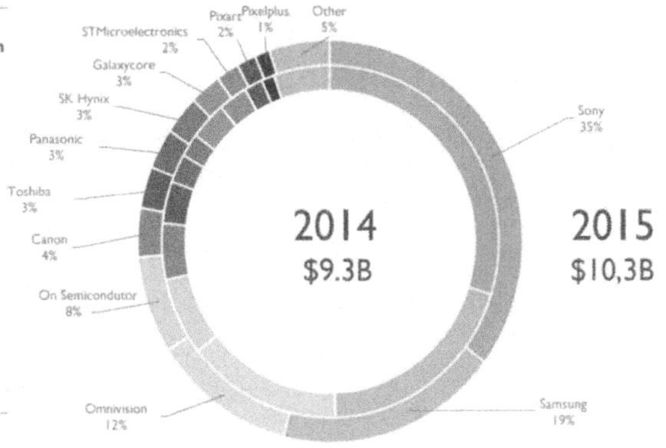

(Yole Développement, June 2016)

Figure 1.8: Image Sensor Player Rankings

Pressure Sensors: There are several types of pressure sensors that measure the pressure of gas or liquid. They vary in technologies used, size, and performance. They convert pressure to an electrical signal that is used for processing. However, they are categorized based on the pressure they measure: pressure relative to a vacuum pressure (absolute pressure sensor), relative to atmospheric pressure (Gauge pressure sensor), relative to another reference pressure (vacuum and sealed pressure sensor), or pressure difference (differential pressure sensors). The use case for pressure sensors varies depending on the market and type of sensor.

The dominant technologies currently used in designing and manufacturing pressure sensors are Micro Electro Mechanical System (MEMS) technologies. MEMS have proved to be the choice for the miniaturization of sensors and specifically pressure sensors.

For example, in the consumer electronics market, MEMS pressure sensors are used in watches and mobile phones for providing absolute pressure in a room (barometric pressure) that is also used for indicating elevation/altitude. These sensors are enabling devices to tell us which floor of a building we are on and are complementary to GPS and other sensors to provide 3-D maps.

In the automotive industry, MEMS pressure sensors are used for measuring engine pressure, gas, oil, and tire pressure. For example, a tire-pressure monitoring system (TPMS) in a car has a pressure sensor in each tire (attached to the valve or as a patch to a tire) and flags the driver if the pressure in each tire is below a safe limit.

The figure 1.9 (Yole development- April 2013) show the market application of the pressure sensor markets and their relative size.

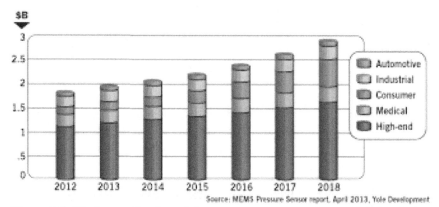

Source: MEMS Pressure Sensor report, April 2013, Yole Development

Figure 1.9: Market application of the pressure sensor markets and their relative size

The top five producers of MEMS pressure sensors are Bosch (Germany), Denso (Japan), Sensata (U.S.), GE sensing (U.S.), NXP Semi (soon to be Qualcomm). The main market that these MEMS sensor companies service is the automotive, industrial, and consumer sensor markets.

Fingerprint Sensors: Fingerprint sensors are starting to emerge as biometric sensors that are used to control access to our computers, phones, building, and other equipment. There are many technologies that have been used to realize fingerprint sensors. These sensors will need to work with the contaminants on our fingers, including sweat, condensation, and lotion/oil, as shown in the figure 1.10 (Source IHS).

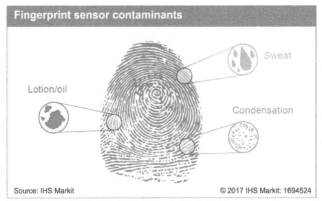

Figure 1.10: Finger Print Sensor

Historically, solving the issues around the detection-accuracy and minimizing the size of these sensors have been a challenge, but the newly adopted technologies (silicon-based or optical-based) and algorithms have solved the issues making these sensors suitable for mass markets to the point that they are used in many of the mobile smartphones. Many smartphones (e.g., iPhone 5s, iPhone 6 Plus, iPhone 6, and 7, Samsung, Huawei, LG, HTC, Huawei) are using these sensors for access point security to the phone and applications that run on the phone in place of passwords (payment, banking, etc.

The market share for 2015 and the major players in this space are shown in the table below.

Company	Revenues	Market Share
Fingerprint Cards	316	50.10%
Synaptics	256	40.60%
Goodix	29.7	4.70%
Silead	7	1.10%
Egis Tech	5.3	0.80%
Elan Microelecronics	3.5	0.60%
FocalTech	3.5	0.60%
ImageMatch Design	1.8	0.30%
Next Biometrics	0.8	0.10%
IDEX ASA	0.03	0.01%
Others	7	1.10%
Total	630.6	100%
Dec. 2015, HIS Technology, Milions $		

In 2015 a total of ~$630 Million worth of fingerprint sensors representing about 545 million units were sold, and this number is forecasted to be $8.9 Billion (about 7.6 Billion units) by 2022, at a CAGR of 19% between 2016 and 2022.

Given the top 3 sensors, their markets, and players, we can see the exponential growth of these sensors that are also forecasted to increase for many years coming. The sensor providers to the companies that integrate these sensors with other electronics, software, and algorithms are taking products to the higher levels of the pyramid of the value chain, going from passive sensors to intelligent senor and sensor nodes that enable many more applications. Next, the architectures that dominate intelligent sensing nodes are covered.

1.4 Internet of Things (IoT) and Intelligent Sensor Nodes (ISN)

It is worthwhile to understand the basic architecture of an intelligent wireless sensor for IoT applications. By the way, the other names that are commonly used for IoT include the Internet of Everything (IoE) and the Internet of Anything (IoX).

The term "Internet of Things" was coined by British entrepreneur Kevin Ashton in 1999. Things connected to the Internet vary in functionality, connectivity, cost, size, power, security, intelligence (brains and analytics), and data bandwidth. These things include but are not limited to supercomputers, roads, buildings, smartphones, tablets, homes, cars, clothing, shoes, appliances, asset trackers, and wearables. Due to the connected things, the Internet as we know it currently will not be the same and will transform into what some call the Internet of Future. Internet of Future is a hybrid of global and local intelligent devices that share data, and the information is provided to everyone at all locations and at all times. The only catch is that you may have to subscribe to it.

Unlike in the past decade, Internet-connected devices are evolving from stand-alone devices to more integrated internet hubs. For example, a smartphone is an ideal hub since it has over ten sensor types, a Global Positioning System (GPS) chip with wired (USB) or wireless connectivity through a Bluetooth Low Energy chip (BLE) and Wi-Fi chip and a 4G cellular chip. Keeping track of activities, position, etc., while being able to communicate with other devices through BLE and to the Internet via Wi-Fi or cellular connection is made possible.

Fundamentally, there are two architectures for the IoT devices that are used in most applications for edge-to-edge connectivity (E2E). Here, by E2E, I refer to the starting point where the actual data is generated, like at a sensor picking up the pulse-rate data. The other edge is where the data is finally analyzed and stored, like a cloud. This is a new paradigm since the source of data is not necessarily located in a device but rather in the environment and has to be sensed and monitored.

So, for E2E connectivity, one edge is the sensor(s), and the other edge is the cloud(s) and/or where data analysis is completed and information is dispersed. Transitioning from the old way of thinking that the personal computer and the mobile phone are the starting node for data, this new paradigm pushes the functionality of the phones, tablets, and laptops to more human-like interface devices and the hub to transfer data to and from the Internet.

Mind you that for machine-to-machine communication to be made through the Internet, a phone, a tablet, or a laptop is not needed.

To better understand the computation and connectivity architecture of IoT devices, a "simple architecture" is used as a reference point. Primarily, the architecture defines components enabling a complete and seamless E2E solution for IoT devices. These components include devices that generate the big data (either raw or partially filtered), hubs that are the center of computation, including connectivity, and then the analytical computation edge with storage in the cloud/fog/edge, as shown in Figure 1.11.

Figure1.11: Showing the components of "simple architecture." Sense Nodes generate the data, passing the information through a middle node such as a hub to the cloud as the other edge node. Note that the arrows indicated the connectivity between each node, which can be wired or wireless and bi-directional.

The second architecture changes the "simple architecture" by making it more application-dependent. In most IoT applications, intelligence and connectivity can be integrated into the sensing nodes.

One can imagine a sensor node that has not only all the sensors but also the embedded processor, as well as one or many wireless connections to the

Internet. In its simplest form, the Intelligent Sensing Node (ISN) requires only one wireless connection to the Internet.

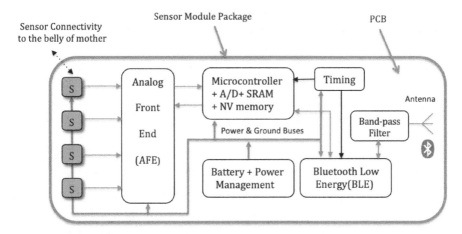

Figure 1.12: ISN consisting of several sensors denoted as 'S', analog front end (AFE), Intelligence with non-volatile memory, Global Positioning System (GPS), memory (MEM), and Antenna.

In Figure 1.12, the sensors monitor and sense parameters in the environment, the Analog Front End (AFE) chips converts sensor data to the microcontroller(s), Global Positioning System (GPS), Non-Volatile Memory (retains data when the battery is out), additional computation memory (MEM) and Antenna is included.

In some applications, ISNs are typically utilized for an asset or activity tracking. For example, when an ISN is placed in a car or truck or attached to an asset, it can track the location of the asset and provide additional information such as temperature, humidity, and the types of motion the asset experiences.

Another example is the smart, intelligent wrist bands that use BLE for connectivity to your phone, have accelerometer/gyroscope, and temperature sensors that track your steps, activity, and the temperature of your body and environment.

Any other type of IoT device architecture can be derived from the two architectures shown. For example, if the ISN node does not have wireless communication directly to the Internet, the connectivity can be done through another physical connection to a hub, then connect to the Internet like the wristbands and smartwatches.

For example, many smartwatches combine sensors and other components of the ISN but are wirelessly connected through BLE to the mobile phone. The mobile phone has the capability of transmitting this data via Wi-Fi or cellular wireless to the Internet (the hub).

To recap, smart sensors that connect the physical world to the Internet are what makes IoT human. They are the eyes and ears of what we sense and are the first edge of the E2E connectivity—sensing anything from sound, vision, chemicals, motions, light intensity, temperature, noise, humidity, pressure, orientation, altitude, and many more.

The intelligent sensor nodes are capable of data interpretation by fusion of sensor data and making sense of it all with respect to the context of the application in conjunction with cloud computing. The next chapters will go through detailed examples.

Chapter 2: ISNs and Our Health

How to monitor, connect, and provide timely healthcare

"I think the biggest problem with healthcare today is not its cost – which is the big problem – but for all that money, it's not an expression of our humanity."

Jonathan Bush (CEO and President of Athena health)

"The best way to find yourself is to lose yourself in the service of others."

Mahatma Gandhi (1869-1948)

2.1 Sensing from Womb to Tomb

Healthcare is a complex topic; the vocabulary used by doctors, patients, researchers, nurses, insurance companies, government entities, and other participants differ from one another, making it difficult to use a single vocabulary that makes sense to all readers. After reviewing the vocabulary used by the World Health Organization (WHO), the Organization for Economic Co-operation and Development (OECD), and The US Department of Health and Human Services (HHS), we use the most common language here and give an overview of what healthcare encompasses and how IoT can have a meaningful impact.

In many developed countries, one of the healthcare metrics (it does not take into account many factors) is the money spent annually per person in that country on health. It is easy to monitor and calculate but does not address the main issues. According to the World Health Organization (WHO), in 2012, over $6.6 trillion US dollars was spent on healthcare, which equates to about $948 per person per year worldwide. The US spent $8,362 per person in 2012. This is not the only metric to consider. The number of metrics to track in healthcare is large, but we believe the following five pillars of healthcare are useful metrics by which IoT can address and accordingly democratize healthcare:

1. Quality of life

2. Enablement of participation in societies in a productive way

3. Accessible healthcare

4. Preventing & detecting conditions that affect health outcomes

5. Life expectancy

We also believe IoT devices will lower the spent-per-patient metric but also address the majority of the five metrics by providing solutions that improve the way the practice of medicine is conducted, the way public health is addressed through regulatory affairs, the way environments impact health, and how we address and manage personal health.

Further, our view is that the above metrics pertain to what is called the three stages of life -- from womb to birth, to adulthood through old age, and to the tomb. Starting with our journey in womb to reaching old age, our personality and health signature are formed through genetics passed on to us. At the same time, they are impacted by the environment that we are exposed to as well as the demands of society, all the way through old age where our physical and mental state of being reach their limit.

The hierarchy of the IoT value chain that shapes the healthcare starts with 1) monitoring and generation of data from sensors, 2) analysis of that data to generate information for diagnosis/prognosis, 3) sharing of the information to help make collaborative decisions with all constituents in the loop through Internet and software, 4) followed with monitoring-analysis-sharing of information real-time in cloud, and lastly, 5) closing the loop real-time with all constituents (patients, doctors, nurses, service providers,..). In some cases, the last step is the delivery of medication to a patient by remote commands.

In order to understand how the IoT value chain improves on the healthcare metrics, real-world examples are warranted and are covered next. There are hundreds of examples, but we have selected some for different life stages and have provided references for the ones we believe are noteworthy. For example, we will cover a specific IoT device for prenatal care, but there exist baby health conditions that can benefit from some form of IoT devices that include but are not limited to Sudden Infant Death Syndrome (SIDS); Abdominal Distension of infants, Excessive Crying; Jaundice; Kernicterus, lethargy, and sleeplessness; respiratory diseases, diaper rash, tear-duct obstruction, and Gut Microbiome Composition. References for these other conditions are given in the appendix.

2.2. ISNs in Prenatal Care (From womb)

Some of the components that form the IoT devices include sensors, computing, connectivity, packaging, integration, and analytics. Combined with an IoT device, we are able to monitor conditions in the womb by devices that are called "Prenatal Monitoring Devices" (PMD). PMDs can be categorized into nice-to-have gadgets like monitoring baby's kicks to devices that monitor critical parameters such as fetal heart rate, blood pressure, blood chemistry, etc., that can help detect serious prenatal issues such as oxygen deprivation to the fetus, Preeclampsia, or Eclampsia. In these cases, data is collected from sensors, then analyzed, and information is provided to the mother, shared between the mother and the doctors to drive the decision-making process -- if all is well or some course of action must be taken.

Traditionally, fetal heart rate monitoring is done at office visit intervals or continuously during the labor to measure the fetus's heart rate and rhythm and is done either internally or externally. Of course, the conversation that goes along between the doctor and the mother during the visit is more important than the monitoring since the evolution of mothers as the optimum biological sensor on how the fetus is doing is far superior to any manmade device. So IoT devices that enable both the monitoring and communications in real-time between mothers and care-providers are preferred.

The fetal heart rate monitoring lets the healthcare provider know how the fetus is doing. The average fetal heart rate is between 110 and 160 beats per minute. It can vary by 5 to 25 beats per minute and be considered normal. The fetal heart rate may change as the fetus responds to conditions in the uterus. For example, when done for the "Non-stress" test, it measures the fetal heart rate in response to fetal movements. An abnormal fetal heart rate may mean that the fetus is not getting enough oxygen or that there are other problems.

During labor, the fetal heart rate is monitored to see the effect on the fetus during uterine contractions, the effect of administered pain medicines or anesthesia during labor on both mother and fetus, the effect of tests done during labor, and the pushing during the second stage of labor. These hospital settings for monitoring fetal health can be applied externally using simple and manual devices such as a fetoscope (a type of stethoscope) or electronic Doppler ultrasound devices using a large machine such as a cardiotocograph (CTG), also known as an Electronic Fetal Monitor (EFM). The electronic equipment provides real-time printed information on paper and is more accurate than a fetoscope.

These machines are large in size, used only in hospital/office settings, and their output is typically in the form of charts on paper. In recent years, IoT-enabled PMDs have made fetal heart rate monitoring simple, small, and inexpensive such that it can be applied and used at home by the mother. An IoT PMD device is simple in that when placed on the mothers' belly, it uses sensors to monitor the baby's heart rate externally. The other advantages of using devices like this are that it connects to a mobile phone and, through an app that runs on the phone, one can see the fetal heart rate on the phone's display, store it for later viewing, or even share the data with the physician; it can even alert the mother and the physician in case of anomalies hence enabling fast and on-time intervention. Further, mothers can socialize/share the data with their loved ones. The hope is that devices like these will further reassure anxious mothers who require monitoring and help cut short the many office visits. Since they are also cost-effective, there will be less hesitation and budget-related second thoughts involved. Doctors appreciate the fact that they could track and diagnose patients remotely, allowing quick intervention and provide easier means to discuss mothers' sense of fetus's well-being with them and others.

Another IoT PMD device exists that keeps track of both the mother and the baby's heart rate on a mobile phone and transfers the data to the mother and doctor on the phone app. Both the mother and baby's heart rate, fetus kicks, position in the womb, electrocardiogram (ECG) is known, and the data is analyzed to predict events during pregnancy with conditional alarm setting for decision and action. An example of such devices is shown in the figure 2.1.

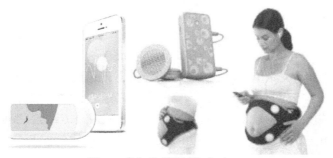

Figure 2.1: IoT PMD device

Fetal heart-rate Fetal and Mother heart-rate & ECG Hand-held sensor node multiple sensors held by a strap.

The IoT PMD devices are in their infancy, but from the onset of their appearance on the market, we can rank their healthcare benefits based on the published articles and social media feedback that the users and doctors have provided to date.

The table below shows the six healthcare metrics and our assessment of those impacted by such PMD IoT devices. A checkmark ✓ indicates that the

IoT device has addressed the metric's needs, and a ✖ mark indicates that the metric's need is not met.

As can be seen, the impact of such IoT devices is considerable. These devices are addressing real issues and further reshape the way medicine is practiced. They are changing where and how the data is collected/analyzed and enable doctor/patient communication in a way more effective than the traditional lab and doctor visits. Healthcare democratization is on the way!

Quality of Life	Enable	Accessible	Prevent & Detect	Life Expectancy
✖	✔	✔	✔	✔

An article by Kaiser Health News shows the effects a home monitoring system had on readmission rates for heart disease patients at Duluth, Minn. - based Essentia Health. The national average rate of readmissions for patients with heart disease is 25%, but after Essentia Health implemented a home monitoring system, the rates of readmission for their heart disease patients fell to a mere 2%. And now that hospitals are being financially penalized for readmissions, home-monitoring systems may offer a solution to avoid those penalties. Furthermore, a large-scale study published in CHEST Journal shows patients in an intensive care unit equipped with telehealth services were discharged from the ICU 20% more quickly and saw a 26% lower mortality rate than patients in a regular ICU.

As Eric Topol states in his book The Patient Will See You Now: The Future of Medicine is in Your Hands, "We are embarking on a time when each individual will have all their own medical data and the computing power to process it in the context of their own world. There will be comprehensive medical information about a person that is eminently accessible, analyzable, and transferable."

By taking a deeper look at the examples, we can see which areas of the IoT value chain it addresses and understand how such IoT devices are architected.

First, the patient(s) is monitored using sensors to generate critical data, and then the data is transferred from sensors to mobile devices using Bluetooth wireless communication. Next, the data is analyzed, and the information for diagnosis/prognosis is provided on the mobile phone, enabling the data sharing between the phones' Wi-Fi or cellular connectivity to the Internet. The shared information then enables the constituents to make an informed decision. Although the data sharing and communications are not done in real-time, the means of closing the loop between the constituents is provided. To better understand the architecture of such devices, it is beneficial to map out the building blocks of these devices and see how they are utilized.

As we indicated in the earlier chapters, there have been many efforts in the standardization of IoT architectures. However, since the world of IoT is in its infancy (it has been at the peak of the Gartner's Hype curve for the last four years), no architectural standards exist that are followed. Due to the fluidity and lack of such standards, many companies have decided to use building blocks, components, and software that already exist and are built into the existing infrastructures (Phones, Wi-Fi, Internet, APIs,...), which are cost-effective and are universally available. As a result, their architecture focuses on the building blocks that differentiate their solution (sensors, APIs, Apps,...) and enables their solution by connecting and utilizing existing infrastructure for a timely market presence.

2.3 ISN device architecture used in Prenatal Care

A high-level data connectivity of a generic PMD is shown in the figure 2.2. The product consists of 1) a sensor node module that embeds several sensors, conditioning circuits, computing, storage, and wireless connectivity to the mobile phone via BLE. 2) The mobile phone is used as a communication hub to the Internet via cellular and/or Wi-Fi. It is also used to display data, do information-extraction from the data, store it locally, and send it to the EHR system that can be viewed by the doctor and other constituents. The figure 2.2

shows how PMD communicated with the phone and makes a connection to the Internet that enables data sharing and communications.

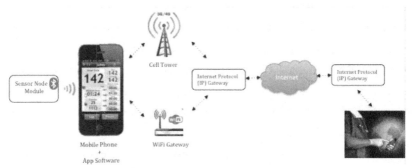

Fig 2.2: Data connectivity from Sensor Node to Phone to Cloud to Doctor

As shown above, other than the sensor node module and the apps that run on the patient and provider's phones/tablets/PC, the rest of the figure's components consist of the existing Internet infrastructure. So, it is worthwhile to look inside the sensor node module and see the block diagram that represents its physical and connected architecture. The figure 2.3 represents such a block diagram.

Sensor Node Module Block Diagram

The sensor node module consists of a printed circuit board (PCB) that integrates and electrically connects all the module components and a module pack that holds the PCB. The PCB components include an analog front chip that conditions and amplifies sensor data and pass that data to the microcontroller.

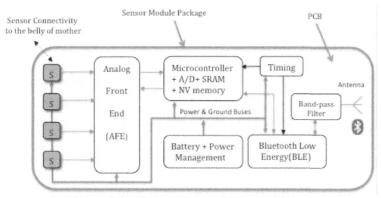

Figure 2.3: SENSOR MODULE PACKAGE

Presented in the figure is a microcontroller that has built-in Analog to Digital converters (A/D) that digitize the sensor(s) data, Static Random Access Memory (SRAM) for temporary storage, and non-volatile (NV) memory for long term storage. The micro does further digital signal processing as it extracts information from the data and stores the data in the SRAM and NV memory.

The embedded software residing in the NV memory controls all the digitization, computation, storage, and communication with the BLE device. The BLE device transmits the information extracted from the sensors to the patient's mobile phone through the Antenna. Other components such as a timing chip provide the clock required to run the microcontroller and the BLE device. The battery and power management block consist of the batteries to power the PMD, and typically power management integrated circuit (PMIC) provides power to all blocks and manages battery life when the device is not in use. Some micros do have PMIC and AFE capabilities integrated as well. There are many companies and choices for each and every chip.

The PCB of the sensor node module is enclosed in a package that either consist of a handheld box that the mother places on different areas of her belly, or the box is embedded in a strap. The strap is used in cases when multiple sensors at different locations (for example, two lead ECGs) are needed. In addition, the strap eliminates the fatigue and the motion artifacts such as unnecessary motions generated when holding the device steady on the body with hands.

In cases where parameters such as ECG are needed, multiple sensors and a single module are located in a strap and leads within the strap carry the sensor signals to a connector at the module. The sensors' location for monitoring ECG signals is very critical, and the strap with the embedded sensors is positioned so that the error for positioning of the sensors does not become an issue.

2.4. ISN in Medical patch device (From Birth to Old Age):

Consulting firm PricewaterhouseCoopers data estimates that nearly 39% of adults would be willing to have an appointment with a physician via smartphone, representing a potential large mobile health (mHealth) market. About half of the doctors surveyed by PwC's Health Research Institute said more than 10% of patient office visits could be replaced by e-visits. About 37% said that one-third of visits could be done virtually. However, mHealth is not only about video chat. It has also become a tool that allows patients to become active players in their treatment when combined with biomedical data sensing and actuation, especially in critical care. In this chapter, we will not cover all the mHealth technologies but a few including, role of IT in healthcare, apps-only software, managing electronic medical records (EMR), telesurgery, and remote doctor visits. References are provided for the reader in the appendix. Here we are giving examples that embed in them the capability of remote monitoring similar to the PMD.

Wearable medical devices use sensors and actuators to enable data collection for analysis, but some of these devices are capable of administering medication via needle-free technologies (NFIT), replacing old hypodermic needles and the bulky large pieces of equipment that have historically been used. These devices are being made in the form of intelligent wearables such as patches, bands, diapers, lab-on-a-chip (LOC) pills, and ingestible pills with varying capabilities and applications. The figure 2.4 shows an example of such devices. Some of the IoT pills are ingestible pills that are being utilized for medication compliance.

Infant Patch Adult Patch LOC Pill

In this section, we have chosen one IoT device example since it has a large impact: it is not invasive (does not require surgery for usage and mostly does not draw blood), is capable of medication delivery in addition to monitoring conditions, and have a distinct IoT architecture that can be used as a platform for other devices. An example is chosen from the IoT devices that fall under the category of intelligent medication patches, also known as medication bandages.

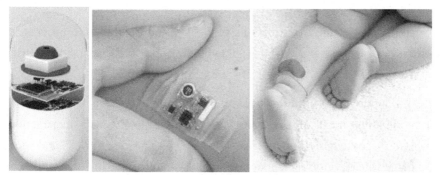

Figure 2.4: Intelligent Medical Patches

There are many examples of medical patches that monitor vital signs and biometric measurements such as heart rate (HR), HR variability, one-Lead ECG, respiratory rate, skin temperature, body moisture, fall detection, sleep cycles, stress level, and activity, including step counting. These patches are well suited for in-patient, outpatient, and home settings where providers can use them to improve the quality of patient care, increase access, and reduce costs. Here we will focus on patches that, in addition to monitoring, are also able to hold a reservoir of medication and deliver medicine to the patient—by the patient, by the health care provider, or by the device if the generation of alerts does not get responses from the patient and the provider.

2.5 Needle-Free devices in Diabetes Care

Diabetes self-care is cumbersome and painful at times since it requires two invasive procedures that are done by the patient or healthcare provider daily. First, it requires the constant need to draw blood for glucose testing and the need for daily insulin shots, and the heightened risk of infection from all that poking. The existing processes of drawing blood and injection are not only painful but also require time for the skin to heal. This process can take up to a week before the same location on the skin can be used again. If a device could monitor and administer the medication without the use of a needle or drawn blood, the patient's comfort is increased, and anxiety of unwanted infections is reduced.

Although attending to infections in the US and much of the progressive world is considered easy, even though uncomfortable, in third world countries, an infection could be a matter of life and death.

Continuous glucose monitors and insulin pumps are today's best options for automating most of the complicated daily blood sugar management processes – but they do not completely remove the need for pricks and shots. Companies are developing technologies that would replace the poke and shots with a patch. These patches have a biosensor that reads blood analytes through the skin without drawing blood.

The products include a handheld micro-abrasion device, ironically also known as microneedle (MN), that removes just enough top-layer skin cells (Epidermis layer) to put the patient's blood chemistry within signal range of a patch-borne biosensor in the dermis layer of skin. Unlike Hypodermic needles, the MN does not penetrate the vein or artery. The sensor collects one to several readings per minute and sends the data wirelessly to a remote monitor, triggering audible alarms when levels go out of the patient's optimal range and tracking glucose levels over time. The figure 2.5 shows the structure of human skin and how the MN devices connect to the dermis layer through the epidermis without drawing blood. There are many types of MN: solid MN that is used for abrasion only and can be removed; coated MN that can deliver drugs then removed; dissolving MN that is coated with drugs and dissolves requiring no removal; and hollow MN that can be used to deliver drugs as needed over long periods of time as shown in Part B of Figure 2.5.

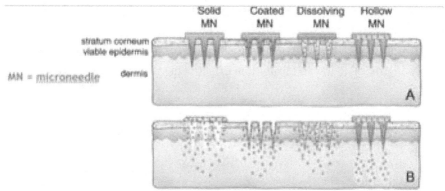

Figure 2.5: Example Microneedle types, how they penetrate the skin and deliver drugs

Needle-free technology (NFIT) is an extremely broad concept, which includes a wide range of drug delivery systems that drive drugs through the skin using any of the forces like Lorentz, shock waves, pressure by gas, or electrophoresis, which propels the drug through the skin, virtually nullifying the use of a hypodermic needle. This technology is not only touted to be beneficial for the pharmaceutical industry but the developing world too; it can be highly useful in mass immunization programs, bypassing the chances of needle stick injuries and avoiding other complications, including those arising due to multiple the use of a single needle. Also, similar to the technologies in nicotine patches that are used for smokers to reduce craving, some drugs can be delivered with an exact dosage by mixing the drug into a bandage adhesive that is time-released by controlling the diaphragm properties that come in contact with skin or electronically controlled releases by use of diaphragms that can change porosity based on the electrical voltage that is applied.

The NFIT devices can be classified based on their working, type of load, drug delivery mechanism, and site of delivery. To administer a stable, safe, and effective dose through NFIT, the drug's sterility, shelf life, and viscosity are the main components of concern. The figure 2.6 shows another example of an NFIT device delivering medication compared to that of a hypodermic needle. As can be seen, the depth of skin penetration for the NFIT device is much less, and hence the required time to heal is drastically improved from days to minutes. Also, the drug's distribution is over a wider surface area, improving the potency and drug time effectiveness.

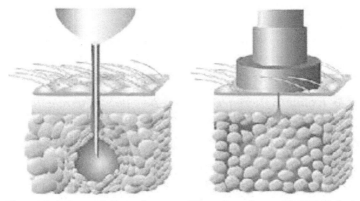

Figure 2.6: Drug delivery by use of Hypodermic needle NFIT device

The IoT glucose patch devices are in their infancy. Still, from the onset of their appearance on the market, we can rank their healthcare benefits based on the published articles and social media feedback that the users and doctors have provided to date.

The table below shows the six healthcare metrics and our assessment of those impacted by such IoT devices. Quality of life metric; Enable being enablement of participation in societies in a productive way; Accessible healthcare; Preventing & detecting conditions that affect health outcomes, and Life expectancy. A checkmark ✓ below the number indicates that the IoT device has addressed the metric's needs, and a ✗ mark below the metric indicates the metric's need is not met.

Quality of Life	Enable	Accessible	Prevent & Detect	Life Expectancy
✔	✔	✔	✔	✔

As can be seen, the impact of such IoT devices is large. These devices are addressing real issues and further reshaping the way critical drugs are delivered to the patient. They are changing where and how the data is collected/analyzed and enable doctor/patient communication differently than the traditional lab and doctor visits. They also enable the patient or the healthcare provider to administer the necessary drugs routinely or, in emergency cases, done automatically when the patient needs more help.

As indicated earlier, no architectural standards exist that are followed when designing such devices, but in the US, the Food and Drug Administration (FDA), and in Europe, CE marking and approval are necessary before such devices are used. In addition, the requirements for the integrity of data, secure connectivity, and reliability take a more central role in their architecture than the PMD devices covered in the last section.

Do you remember the 2007 incident where US Vice President Dick Cheney ordered that wireless features be disabled on his defibrillator and the fact that the popular US TV thriller Homeland took inspiration from Cheney and actually depicted the murder of a vice-president via a hacking attack on his pacemaker?

The importance of securing these devices can only be more socialized by the examples that the Federal Bureau of Investigation (FBI) alerts post on the association for the advancement of medical instruments (AAMI) website. These alerts are specific to IoT medical devices, and their recommendations are:

- Isolate IoT devices on their own protected networks.

- Disable UPnP [Universal Plug and Play protocol] on routers.

- Purchase IoT devices from manufacturers with a track record of providing secure devices.

- When available, update IoT devices with security patches.

- Be aware of the capabilities of the devices and appliances installed in homes and businesses.

- If a device comes with a default password or an open Wi-Fi connection, consumers should change the password and only allow it to operate on a home network with a secured Wi-Fi router.

- Use current best practices when connecting IoT devices to wireless networks and when connecting remotely to an IoT device.

- Be informed about the capabilities of any medical devices prescribed for at-home use.

- If the device is capable of remote operation or transmission of data, it could be a target for a malicious actor. Ensure all default passwords are changed to strong passwords. Do not use the default password determined by the device manufacturer.

In the next section, we will explore examples of the architecture of such devices with a detailed block diagram that includes data security.

2.6 Glucose patch device architecture

There are several distinct architectural and physical differences between patches and PMDs. The first one being the way they are connected to the Internet through what is called Personal Area Networks (PAN) that are designed for a single person connected to the internet bridge. PAN can be used for communication amongst the personal devices themselves (interpersonal communication) or for connecting to a higher-level network and the Internet where one "master" device takes up the role of an internet router. The operating range of PAN is limited by design to shorter distances (between 1 foot to 100 feet) and has encryption/decryption of data and passwords for secure transmission and reception of data. If the PAN has wireless communication, it is called a Wireless Body Area Network (WBAN). WBAN connects independent nodes (e.g., sensors and actuators) that are situated in the clothes, on the body, in the body, or under the skin of a person.

The network typically expands over the whole human body, and the nodes are connected through a wireless communication channel. Second is the use of security encryption/decryption technologies that these devices adopt in order to make health data, passwords, and communications more secure from hackers or unwanted access. In addition, to secure the PAN interface, each device has embedded security hardware and software.

The third is the MN and the medicine reservoir that enables drug delivery through the skin. Fourth is the battery technology that is adopted in patches and is becoming very low profile; it consists of a thin film or polymer batteries. Last is the Flexible Printed Circuit Board (FPCB) that conforms to the shape of the skin, connects all the components, and holds the medication reservoir. The connectivity of the patient to a PAN is through a mobile phone or hub with secured software all the way to the hospital with notification devices, as it is shown in the figure 2.7. Multiple users can use the gateway, but the PAN is used once per person. As can be seen, other than the sensors, secured PAN, all the other components for the IoT connected devices are the existing infrastructures similar to the PMD connected device described before.

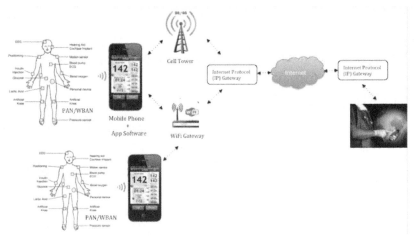

Figure 2.7 : IoT devices and the human body

It is worthwhile to look at the block diagram showing the components, their interconnection that forms the architecture of the NFIT patch device. The patch consists of a Flexible Printed Circuit Board (FPCB) that integrates and electrically connects all the components in the patch, and a flexible substrate with adhesives that are skin-friendly, attach to the skin and hold the FPCB. The components on FPCB include the sensor(s), the microneedle(s), two medication dosage control, one of which is used as a fail/safe control in case of emergencies, the analog front end that conditions and amplifies sensor data and passed that data to the microcontroller, a chip that does encryption and decryption of data, and passwords, timing clock generator, wireless connectivity for PAN (we show a BLE device, but it can be any other standard for short-distance communication protocol).

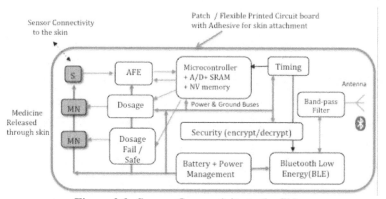

Figure 2.8: Sensor Connectivity to the Skin

A microcontroller that has built-in Analog to Digital converters (A/D) that digitizes the sensor(s) data, Static Random Access Memory (SRAM) for temporary storage, and non-volatile (NV) memory for long term storage, does further digital signal processing, extracts information from the data and stores the data in the SRAM and NV memory. Some microcontrollers also come with AFE capability. In that case, the AFE block is integrated into the microcontroller. The embedded software residing in the NV memory controls all the digitization, computation, storage, and communication with the BLE device. The BLE device transmits the information extracted from the sensor(s) to the mobile phone of the patient through the Antenna; it also receives commands from the patient or care-provider to deliver medication which is transferred to the microcontroller; a decision on validation is made, and that command is sent to the dosage control block. Other components, such as the timing chip, provide the clock required to run the microcontroller and the BLE device. The dosage blocks receive commands from the micro and deliver the correct dosage of medication that is in a reservoir placed near or is part of the microneedle(s). The battery and power management block consists of the batteries to power the patch, and typically a Power Management IC (PMIC) that provides power to all blocks and manages battery life when the device is not in use. The FPCB of the patch conforms to the skin, makes sensor and MN contact to the skin, holds the medicine reservoir, and is inert to the skin.

2.7 Putting it all together:

We have covered real-life examples of how IoT devices address the five pillars of healthcare and potential cost benefits. We have shown that quality of life is improved by these devices, participation in societies in a productive way is enabled, accessible healthcare is attained, prevention and detection of conditions that affect health outcomes can be achieved, and more importantly, statistically speaking, the life expectancy of humans is improved. Further lower costs are attained by reducing doctor, hospital, and lab visits.

We have also shown that some of these IoT devices provide and complete the full circle on the hierarchy of value chain, including monitoring and generation of data from sensors, then analysis of data to generate information for diagnosis/prognosis, sharing of the information to help make collaborative decisions with all constituents in the loop through Internet and software, followed with monitoring-analysis-sharing of information real-time in cloud, and ends with closing the loop real-time with all constituents (patients, doctors, nurses, service providers,..).

Putting it all together, the future of health care can be summarized in the example picture shown below. From Pill Cameras to patches to implantable devices and other wearables that sense our health status communicate it through PAN to the cloud, storing the data, sharing it with the care providers in real-time with control and feedback achieves all five pillars of health care and potentially making the cost of healthcare lower.

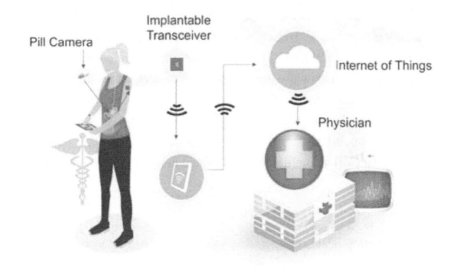

Figure 2.9: Health Care and IoT

In many medical conditions, the environment also impacts our health, so it is common to include environmental sensors that are highly integrated into these devices. Many sensor companies now have environmental sensors that can measure the temperature, extract pollution levels by measuring the ratio of gases such as oxygen, carbon dioxide, carbon monoxide, and other Volatile Organic Chemicals (VOCs) in the environment. In addition, they incorporate pressure sensors to measure ambient pressure, accelerometers/gyroscopes to measure motion, and provide all this data along with the patients' medical data into the cloud.

Imagine how powerful our analysis of data would be if we do health and environmental monitoring of not only a single individual but collect such data from populations in specific areas, cities, countries, and continents. According to the World Health Organization, 80% of chronic disease deaths occur in low-to-mid income countries.

Being able to have access to such a vast amount of data to extract useful information would not have been possible before these IoT devices. Being able to correlate the regional and global environmental data on local and global health is a new white space that we are entering and enabling. The emergent technology paradigms of Big Data, Internet of Things (IoT), and Complex Event Processing (CEP) have the potential not only to deal with pain areas of the healthcare domain but also to redefine healthcare offerings for a healthcare system that builds upon integration of Big Data, CEP, and IoT. Event processing for healthcare applications, specifically IoT-based applications that combine real-time data streams from medical devices with other patient and community data, offers multiple benefits, including:

- • Filtering out "noise" or normal, uninteresting, patient readings to help in many scenarios, including alarm fatigue [SEP]

- • Alerting healthcare personnel to alarming patient medical readings and trends

- • Creating closed feedback loops with devices that monitor a patient's health and devices that administer medication or treatment based on those readings, such as infusion pumps and patches

- • Keeping family members and friends in the loop for occasions such as lack of patient movement or mobility, detection of medication non-adherence, proactive emergency detection, communication, and so on.

- • Making real-time correlations with environmental and population health statistics and community health data. For example, to detect cancer clusters, spikes in asthma attacks in a community, detection of mold in a home and mold-related illness, and so on.

In short, CEP can be used to predict and administer critical care as needed before it is too late, all based on a doctor's prescribed care. The era of complex event processing, especially within IoT solutions, is upon us and can be leveraged to accelerate safe innovation in many areas of healthcare technology. We are taking the first step of a thousand-mile journey! [SEP]

Chapter 3: Sensors and Our Food

"Businesses must evolve over time, and as more information on food safety becomes available to the public, those that pro-act to new advances are going to be in a better position."

Roger Berkowitz, President, and CEO, Legal Sea Foods

"Let food be thy medicine and medicine be thy food."

Hippocrates, Physician

3.1 From Growth to Consumption

Food safety and quality are topics that have been highly debatable for the past decades, resulting in food policy within the industry and more funded research. This complex issue can be viewed as two topics that have driven the debates. First, a variety of food scares has directed public attention to food safety issues. Second, the general public has become interested and often critical with regard to certain ways of food production as they dictate food quality. Consumers in developed countries have become more demanding, more critical, and more fragmented in their food choices, leading to situations where quality differentiation of food products has become necessary in order to satisfy consumers.

In the following sections, food safety, quality, and production are reviewed to show the gaps to an ideal state of knowing the quality and safety of our food from Growth to Consumption (G2C) before consumption and how intelligent sensing technology can get us closer to the ideal state. This is a long-term journey that we are taking from not knowing and having sporadic data about our food safety and quality to sensing, monitoring, analyzing, and having real-time data of our food hence moving up the pyramid of value chains.

Food Safety

Center for Disease Control (CDC) estimates that each year roughly 1 in 6 Americans (or 48 million people) gets sick, 128 thousand are hospitalized, and 3,000 dies of foodborne diseases. The ingesting of bacteria, fungi, parasites, viruses, toxins causes foodborne illnesses or other harmful substances in contaminated food. Foodborne illness is a serious problem. The top five pathogens causing 91% of the foodborne illnesses include Norovirus (58%), Salmonella (11%), Clostridium perfringens (10%), Campylobacter spp (9%), Staphylococcus aureus (3%), and E.Coli (<1%). In 2015, CDC announced the Chipotle-Linked E. Coli outbreak case count to be 52. The final actual number is larger than this, with recurring cases in 2017.

On January 4, 2011, the Food and Drug Administration (FDA) in the US passed the Food Safety Modernization Act (FSMA), the most sweeping reform of our food safety laws in more than 70 years. It aims to ensure the US food supply is safe by shifting the focus from responding to contamination to prevention. The main guidelines for prevention include keeping clean (wash your hands), separating raw and cooked food, cooking thoroughly, keeping food at safe temperatures, and using safe water and materials. Although these are good guidelines, they are just that and not adequate for true prevention. Sensing, monitoring, analyzing, sharing of data, and real-time prevention will help us get closer to true prevention when combined with the existing guidelines. Here lies the opportunity for intelligent sensors to help close the loop. The pyramid of the value chain described in chapter 1 is yet to be realized and implemented for prevention regarding foodborne illnesses.

Imagine a world where sensors can detect, notify, and share the data on the pathogens in raw food, water as well as cooked food prior to consumption. We cannot only improve the quality of life for those who get impacted (over 48 million in the US) but also reduce the cost of healthcare arising from foodborne illness currently estimated to $52B to $78B alone in the US.

Based on the technologies, laws, regulations, and applications, I estimate that such real-time preventive systems can be in place by 2030. Depending on the food types and location from harvest to store and to home, the sensors, packaging, and detection types vary. In this chapter example of ISNs that will pave the path to improve food safety is given.

Food Quality

Food quality is the characteristics of food that include external factors such as appearance (size, shape, color, gloss, and consistency), texture, flavor, and factors such as federal grade standards and nutrients for growing/raising food (chemical, physical, microbial).

The quality of food starts with the raw material, ingredients, followed by sanitation standards in producing food, transporting/storage of food, and preparation of food prior to consumption. In agriculture, harvest quality is dictated by the quality of seeds, water, nutrients, and the soil used in producing food. In animal farming, the quality is determined by the lack of animal byproducts, lack of antibiotics, lack of hormone levels in feeds, and lack of crates, cages, or tethers in the life span of animals. In transportation and storage, quality is dictated by the speed of transportation, refrigeration, and handling of the food at the right temperature as some of the parameters that determine food quality.

Although many quality standards are also dictated in labeling, handling, storage, preparation, and traceability, we are still doubtful about the food we buy. Why is that? It is the lack of data and information that we need to make informative and quick decisions about what we eat and when we eat it. What ISNs can do is close the loop from farm to home and provide the needed information in a timely manner to us for making an informed decision.

The following sections give specific examples of how we can improve the quality of our food using ISNs.

3.2 Sensors applications in agriculture (from the farm)

Agricultural sensors are typically used to monitor crops health, soil health, and quality of crops. These sensors can be located in soil, on-crop, on drones/airplanes, or on satellites for remote sensing. The placement of sensors determines the granularity of data and information we receive. The closer the sensor to the food source, the more quantity of sensors and more granular information is provided, whereas the farther away from the sensor from the food source, the fewer nodes and less granular data is needed. Examples of these scenarios are given in the following sections.

3.2.1 ISNs located in soil

As the name implies, in this class of ISNs, the sensors are embedded in the soil and making contact with the soil at different depths. United States Department of Agriculture (USDA) defines the following indicators examples that are related to soil quality, in addition to guidelines on how to assess soil quality.

- Soil organic matter => nutrient retention; soil fertility; soil structure; soil stability; and soil erosion

- Physical: bulk density, infiltration, soil structure, and macrospores, soil depth, and water holding capacity => retention and transport of water and nutrients; habitat for soil microbes; an estimate of crop productivity potential; compaction, plow pan, water movement; porosity; and workability

- Chemical: electrical conductivity, reactive carbon, soil nitrate, soil pH, and extractable phosphorus and potassium => biological and chemical activity thresholds; plant and microbial activity thresholds; and plant available nutrients and potential for N and P loss

- Biological: earthworms, microbial biomass C and N, particulate organic matter, potentially mineralizable N, soil enzymes, soil respiration, and total organic carbon => microbial catalytic potential and repository for C and N; soil productivity and N supplying potential; and microbial activity measure

An example of types of measurements, frequency of data collection, and location of in-soil sensors is shown below (Source: neon)

Sensor Type	Sensor Location	Measured Frequency
Carbon Dioxide	down to feet (s) below the soil	10X /second
Heat Flux	down to a foot below the soil	10X /second
Line Quantum sensor	On the surface	1X /second
Net Radiometer	On the surface	1X /second
Solid Moisture	~ 7 feet below the soil	1X /second
Soil Temperature	~ 7 feet below the soil	2X /second
Throughfall	Surface	

Carbon Dioxide (CO2) sensors

CO2 sensors are used mainly to assess Soil respiration (Plant Roots, the rhizosphere, microbes, and fauna) that release carbon from the soil in the form of CO_2. CO_2 is acquired from the atmosphere and converted into organic compounds in the process of photosynthesis. Plants use these organic compounds to build structural components or respire them to release energy. When plant respiration occurs below-ground in the roots, it adds to soil respiration. Over time, plant structural components are consumed by heterotrophs. This heterotrophic consumption releases CO_2, and when this CO_2 is released by below-ground organisms, it is considered soil respiration. In soil, CO_2 respiration reflects the quantity and quality of organic humus, which is essential to sustainable fertility. In compost, it reflects the age and safety of the material. The higher the level of biological activity of soil humus, the healthier soil is, and the more nutrients - like nitrogen and phosphorus - it will provide naturally for growing plants. In addition, microbial-rich humus in soil and compost is a key to excellent soil-tilth and promotes natural resistance to soil-borne plant diseases. Combining this information with other indicators of soil health, such as organic matter, we can begin to get a clear indication of how healthy soil is and its ability to release nitrogen.

Soil Heat Flux Sensors

Heat flux sensors measure the rate of energy transferred through a given surface. The sensors can be several thermocouples whose measurements are averaged, a thermopile, or a thermopile with a film heater. Soil heat flux tells us how much energy is stored in the soil as a function of time. Typically two or three sensors are buried in the ground at a depth of around 1 to 12 inches below the surface. Soil differences are observed by recording the amount of energy transfer capability through the soil. If you walk across your lawn, there are spots where the grass is thick and other spots where it is thin, and you can see bare soil. This is an indication of the variation of soil energy storage capability in different areas of the soil. Similar to lawns, there are fields of corn, wheat, and pasture that have large spots (acres) where biomass is thick and other large spots (acres) where it is thin. The difference is usually due to differences in soil fertility (the relative ability of a soil to supply the nutrients essential to plant growth).

Line Quantum Sensors

Line Quantum sensors measure Photosynthetically Active Radiation (PAR) under a canopy or in the field. PAR light lies in the range 400-700nm wavelength and indicates how much of the sunlight is present in the field that can be used for photosynthesis by the plants. The measured light is in units of Photosynthetic Photon Flux Density (PPFD), which is expressed as μmol s-1 m-2. If a field is covered with overhang trees or another object, or if a canopy is selected to raise plants, these sensors help to decide whether a given location receives adequate sun for a given crop or plant.

Net Radiometer Sensors

Net radiation is the balance between incoming radiation from the sun and sky and outgoing radiation from the ground. Short-wave radiation of 0.3 to 3 μm wavelength reaches the Earth's surface, where some of it is reflected, and the rest of the energy is absorbed by the surface. Incoming long-wave Far Infrared (FIR) radiation from 4.5 to more than 40 μm is also absorbed by the surface, which heats up and emits FIR back to the sky. The main applications for net radiometers are in agro-meteorology, in particular for the study of evapotranspiration. This is the sum of evaporation and plant transpiration from the Earth's land and ocean surface to the atmosphere. Evaporation accounts for the movement of water to the air from sources such as the soil, canopy interception, and waterbodies.

Soil Moisture and Temperature Sensors

A soil moisture sensor combined with a temperature sensor is placed at different depths in soil and measures how much moisture, heat/cold is absorbed in the root system of the plant, informing us of the moisture content and enabling the automatic watering system to work more efficiently. This improves the productivity and quality of the food. These sensors are used in sports turf, residential, landscaping, and ground care to reduce and optimized water consumption. The soil moisture sensors fall under two classes of sensors, volumetric soil moisture sensors, and soil water potential sensors, and at times are not combined with temperature sensors. These sensors are further categorized based on the technology that they utilize to measure moisture (gypsum block, GMS, tensiometer, probes, capacitance sensor, and TDT). The total market for these sensors is **estimated** to reach $206.2 Million by 2020, at a CAGR of 16.2% between 2015 and 2020.

The major players in this market include, **Acclima, AquaCheck, Decagon Devices, Inc., The Toro Company, IRROMETER Company, Inc.,** Delta-T Devices Ltd. (UK), Sentek Pty. Ltd. (Australia), AquaCheck (Pvt) Ltd. (South Africa), Acclima, Inc. (US), Sentek, The Toro, and Stevens Water Monitoring Systems, Inc. (US) among other upcoming and startup companies.

Soil moisture sensors are placed at different depths, and the depth is determined based on changes in soil texture, structure, or color in the profile. A soil moisture sensor that also measures electro conductivity could also be used at deeper depth to evaluate the movement and accumulation of salts.

Having the moisture data enables the irrigation system to optimize water usage and water the plant when needed, not over or under water. An example of such a sensor used for automatic vineyard irrigation management is shown **below.**

Figure 3.1 Irragation sensors

Sensor placement showing Depth at varying soil textures

These types of sensors are typically wired or wireless and provide real-time data to the irrigation systems to control water release. Example data from soil moisture sensors at different depths are shown below (data is taken by the University of Wisconsin – Civil and Environment Engineering Department). The four types of vegetation, soil type, the depth of sensor, and % water content over time with the relative location of the plants are given.

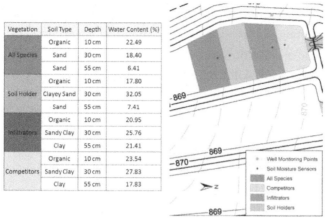

Vegetation	Soil Type	Depth	Water Content (%)
All Species	Organic	10 cm	22.49
	Sand	30 cm	18.40
	Sand	55 cm	6.41
Soil Holder	Organic	10 cm	17.80
	Clayey Sand	30 cm	32.05
	Sand	55 cm	7.41
Infiltrators	Organic	10 cm	20.95
	Sandy Clay	30 cm	25.76
	Clay	55 cm	21.41
Competitors	Organic	10 cm	23.54
	Sandy Clay	30 cm	27.83
	Clay	55 cm	17.83

Figure 3.2: The four types of vegetation

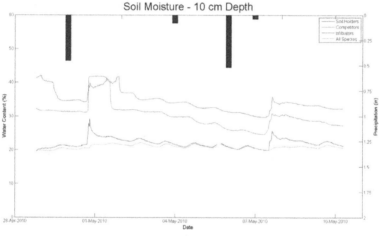

Figure 3.3: An example of real-time at a given soil depth data from the last figure.

The most accurate soil probe will be fully surrounded by the soil, with no gaps or air holes between the probe and the soil. The probe then sends electrical signals into the soil, measures the responses, and relays this information to a data collection device known as a data logger. It then transmits the data to a central computing system.

ARTIFICIAL SENSORS

What makes the information from soil sensors so valuable is that you are able to install multiple probes in the same area, each one buried below the last. This gives you an idea of what water is doing as it moves down through the different layers of soil (known as soil horizons).

The table below shows the six-quality metrics and our assessment of the ones that are impacted by such devices. Quality of life metric; Enable being enablement of participation in societies in a productive way; Accessible healthcare; Preventing & detecting conditions that affect health outcomes, and Life expectancy. A checkmark ✓ below the number indicates that the IoT device has addressed the metric's needs, and a ✗ mark below the metric indicates the metric's need is not met.

Quality of Life	Enable	Accessible	Prevent & Detect	Life Expectancy
✓	✓	✓	✓	✓

3.2.2 ISNs located on-crop

As the name implies, in this class of IoT devices, the sensors are placed on or near the crop to mainly monitor moisture, color, and temperature and extract the health of the crop.

ISNs can be integrated into a number of 'platforms' – a term that essentially means mediums in which sensors can be placed to achieve varying purposes. Sensors integrated into an on-ground platform would mean placing a robot with ISNs installed in it near or in the field to yield objective information from the crops. The biometric traits of a crop can be studied "in a tri-level division of crop features depending on the focus of interest being at soil level, plant level, or produce-level." (Saiz-Rubio and Rovira-Mas). The goal is to get a larger number of non-invasive sensors that can perform the function of prediction and data collection with accuracy and distance. Although multispectral and hyperspectral technologies are making progress, most of the sensors available are operative from a distance from the target, i.e., crops. The closest they can be placed is within 2 meters since they are mounted on mobile robots, as shown in Figure 3.4.

These are also called Ground Autonomous Systems, and the method is called Proximal Sensing. Ground-based platforms or ground vehicles acquire accurate data, the accuracy being inversely proportional to the distance. They are capable of taking one or more samples per meter, further depending on the specifications of the sensors being implemented. When active sensors are used, weather conditions such as strong sunlight or low illumination are not a serious problem anymore, and, in the case of on-the-fly processing, real-time applications are possible, as spraying weeds with the detection of the pest.

Figure 3.4: A prototype surveillance field robot. AdaptoveAgroTech.com

Field Scouting Robots or Unmanned Ground Vehicles (UGVs) have sensors for recording:

- Plant status

- Disease incidence

- Infestations affecting crop growth

When merged with other robots, they also help in:

- Weed control

- Field Scouting

- Harvesting

In the case of harvesting, for example, the sensing mechanism has to identify the ripeness of fruits in the presence of various disturbances in an unpredicted heterogeneous environment, while the actuation mechanism should perform motion and path planning to navigate inside the plant system or tree canopy with minimum collisions for grasping and removing the soft fruit delicately (Shamshiri et al.). Many vineyards managing UGVs are being developed to cut off on manual labor and make management easy. Some UGVs of this kind include VineRobot, Vinbot, GRAPE, and VineScout.

(a)   (b)

Figure 3.5. Version II (2019) of VineScout autonomous robot: Front (a) and rear (b

It monitors vine canopies through its fast sensors; it is cost-efficient, and the integration of its software with the ground-truth validation may have been a challenge, but it makes the task ten times easier nonetheless. Such on-ground sensors can also make fertilizer recommendations. Certain reflectance sensors are also programmed to predict yield potential and crop nitrogen response. When this crop sensor information is merged with soil moisture information, more holistic calibrations are made possible for future usage to produce more crops for the increasing population.

The table below shows the six metrics and our assessment of the ones that are impacted by such devices. Quality of life metric; Enable being enablement of participation in societies in a productive way; Accessible healthcare; Preventing & detecting conditions that affect health outcomes, and Life expectancy. A checkmark ✓ below the number indicates that the IoT device has addressed the metric's needs, and a ✗ mark below the metric indicates the metric's need is not met.

Quality of Life	Enable	Accessible	Prevent & Detect	Life Expectancy
✓	✓	✓	✓	✓

3.2.3 ISNs on drones /airplanes for crop/soil

As the name implies, in this class of IoT devices, the sensors are placed on a drone or airplane to remotely monitor large field areas for moisture, color, and temperature of the crop and the soil surface.

Aircraft and drones are a meter to kilometers away from the crop, bringing in the possibility of less accurate but much larger area coverage, and are compensated by their ease of application and data collection. Aerial and terrestrial data sources are categorized under Proximal Sensing, which can help us understand plants and trees' physiology better than satellites that are kilometers away. As Saiz-Rubio and Rovira-Mas explain, "Unmanned aerial vehicles (UAV) and remotely-piloted aircraft (RPA) can basically be of two kinds: Fixed-wing aircraft and multirotor aircraft." The latter are more stable as they can vertically land and take off as well, while the former can cover more area and carry larger loads as compared to the other. Rotary-wing UAVs are, although, are cheaper and more robust, while Fixed-wingers are more expensive and easily breakable and prone to accidents.

Compared to Field Scouting Robots or UGVs, UAVs can collect data from places inaccessible on-ground sensors. Conventional equipment cannot go into the nooks and corners, which are accessible from the top through an over-head drone or aircraft. While they have this advantage, their route needs to be planned beforehand, and there are certain vision problems faced by them depending upon their surroundings, i.e., dust or fog. Certain machine vision applications may require flying at midday to avoid vegetation shadows on the ground, causing errors with imagery data. Moreover, due to the lack of limits on the load that can be carried by drones, they cannot be installed with a large number of sensors. Strong winds are also not their best companions, but they win in terms of accessibility and pace of the process.

Drones do not only sense the conditions on the ground but are also capable of spraying fluids in required quantities; smart agriculture companies like Intelligent Barn provide such gadgets. They serve as plant-protection machines that come with portable medicine containers. The data they pick up can is transferred on an on-ground configured device in real-time. The data collected by UAVs is high in Spatio-temporal resolution and is decluttered in real-time to provide valuable information on the nature of crops.

Figure 3.6: A micro-UAV measuring crop height

Obtaining estimates of crop height helps characterize the growth rate and health of plants, as well as determining the impact of drought, genetic variations, and other environmental stresses on the crops. Further crop and plant treatments may be devised depending upon this data collected through the sensors in the drones. The manual equipment currently used to perform the same function is time-consuming and damaging to the crops; it is efficient to replace them with UAVs requiring minimum time and invasion. The system studied by Anthony et al. (2014) utilizes low-cost sensors and a UAV platform to reduce costs and operator risks, increase operating ease, and be highly portable. The system is built using a commercial micro-UAV, laser scanner, barometer, IMU, and a GPS receiver to effectively operate the UAV over crops, estimate the UAV altitude, and accurately measure the crop's height by applying a series of onboard filters and transformations.

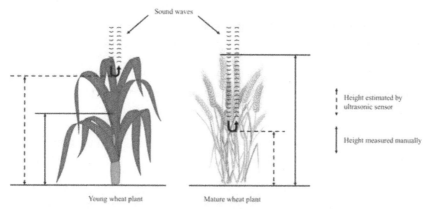

Figure 3.7: Source: mdpi.com

An article on Wheat Height Estimation Using LiDAR in Comparison to Ultrasonic Sensor and UAS by Yuan et al. founded that where "Canopy heights of 100 wheat plots were estimated five times during a season by the ground phenotyping system and an unmanned aircraft system (UAS), the results were compared to manual measurements. Overall, LiDAR [Light Detection and Ranging] provided the best results," hence proving that sensors work better than manual data collectors and would be befitted from if aptly applied.

The table below shows the six metrics and our assessment of the ones that are impacted by such devices. Quality of life metric; Enable being enablement of participation in societies in a productive way; Accessible healthcare; Preventing & detecting conditions that affect health outcomes, and Life expectancy. A checkmark ✔ below the number indicates that the IoT device has addressed the metric's needs, and a ✗ mark below the metric indicates the metric's need is not met.

Quality of Life	Enable	Accessible	Prevent & Detect	Life Expectancy
✔	✔	✔	✔	✔

3.2.4 ISNs located on satellites for crop/soil

As the name implies, in this class of IoT devices, the ISNs are placed on Satellites or high-altitude airplanes for remote monitoring of large field areas for parameters such as crop moisture, crop color, crop temperature, and the soil surface.

Satellites are an effective way of remote sensing farming conditions. In order to do both, enhance food production and promote environmental sustainability, the variability of the ground and soil needs to be studied. This can be done through geostatistics that helps us understand environmental properties and produce maps that can help us understand land placement. Both topographic attributes and electromagnetic information is used to improve the estimation of soil texture and other soil fertility features.

A working example of such systems include the American Landsat satellites (eight satellites take spectral data from the Earth each 16 to 18 days), the European Sentinel 2 satellite system (it provides multispectral data at 10 m pixel resolution for NDVI—Normalized Difference Vegetation Index—imagery, soil, and water cover every ten days), the RapidEye constellation (five satellites provide multispectral RGB imagery, as well as red-edge and NIR bands at 5 m resolution), the GeoEye-1 system (captures multispectral RGB data and NIR data at a 1.84 m resolution), and the WorldView-3 (collects multispectral data from the RGB bands including the red-edge, two NIR bands, and 8 SWIR bands with a resolution of 1.24 m at nadir).

Another example of a working satellite is NASA's own MARS (Monitoring Agricultural Resources); it is a crop monitoring service that helps them understand the levels of global food security. It picks up on crop and yield information to help target better inputs and plan ahead for harvests. For example, using infrared imagery captured by Landsat satellites and publicly available on the Internet through Google Earth Engine, EEFlux can quickly create maps of evapotranspiration, a way to measure how much water is being used is shown in the map provided by NASA. In this map of a region in California, dark blue and dark green represent higher levels of evapotranspiration, while light brown represents low levels.

Figure 3.8: Levels of evapotranspiration

The sensors used in remote sensing satellites for smart agriculture are of spectral, electrical, electromagnetic, or radiometric kinds to study the Earth from above. They help study soil attributes using thermal readings that make judgment calls easier for agriculture scientists. Prediction and information of climate conditions also help in timely decisions and forecasting.

The table below shows the six metrics and our assessment of the ones impacted by such devices. These are respectively: Quality of life metric; Enable being enablement of participation in societies in a productive way; Accessible healthcare; Preventing & detecting conditions that affect health outcomes, and life expectancy. A checkmark ✔ below the number indicates that the IoT device has addressed the metric's needs, and a ✘ mark below the metric indicates the metric's need is not met.

Quality of Life	Enable	Accessible	Prevent & Detect	Life Expectancy
✔	✔	✔	✔	✔

3.3 ISNs in food transportation

According to the Centers for Disease Control and Prevention (CDC), one in six Americans contracts a foodborne illness annually. Of these, 128,000 require urgent care, and around 3,000 meet fatal ends[1]. These foodborne illnesses primarily result from food products gone bad in the process of transportation from source and packaging to market. Food safety during transport is a big concern for companies, and to ensure both the good health of consumers and the reputation of their name, measures are taken to improve and uphold standards of food as it goes through the many steps before reaching the consumer.

---

[1]D. K. Singh, R. Desai, N. Walde, P. B. Karandikar, "Nano warehouse: A New Concept for Grain Storage in India," 2014 International Conference on Green Computing Communication and Electrical Engineering, 2014.

Understanding Food Safety Management Systems Certification Standards

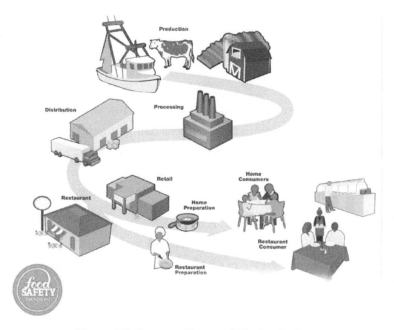

Figure 3.9 Source: Glenwood Technologies

According to a recent report from the Food Marketing Institute, fresh groceries is the biggest category of consumer spending in retail, with consumers projected to spend upward of $100 billion by 2025 on it. Food shipments, including those of fresh produce, travel a long distance, going through different environments with various pollutants and temperature variants. It is in this area where IoT comes into play. Cold chain food processing is made possible through the industrial technologies of sensing not only the temperature of the food being transported to ensure it is optimum but also the vibration, humidity, and light it receives.

Previously, the drivers of food shipments visually ensured the functioning of a thermometer placed in packages on their breaks, but this was time-consuming and prone to error. With cold chain technology, the following conditions can be read in real-time to ensure quality on different fronts:

Temperature: It may be the easiest factor to track, but little fluctuations missed can cause damage to foods like vegetables, bringing down their quality. Therefore, ISNs installed in shipment vessels enable people to keep a check on temperature in real-time, not letting them miss the slightest temperature change.

Humidity: For fruits and vegetables, it is just as important to track humidity as it is to track temperature. These natural foods change in taste and face if put through conditions other than ones ideal for keeping them fresh. ISNs read environmental humidity to ensure it matches the setting required for fruits and vegetables.

Vibrations: For foods like fruits and other soft items, too many vibrations mean bruising and damage to their quality. If the placement of a product in transport is kept in check to avoid high vibrations, quality is assured.

While sensors provide data on the overall conditions of the shipment, whether it is of fresh produce or other products, they can be placed individually on cartons and pallets to keep a check. Sensors work well with all ranges of the environment, from less-than-truckload (LTL) to big, enormous shipments weighing tons.

The supply chain of food cargo is made easier to manage through the use of intelligent sensors. Food production, processing, packaging, distribution, and storage are of great concern. The information provided by smart sensors in these processes is uninterrupted, and this allows for actions to be taken mid-transit if the conditions fluctuate, safeguarding the buyers' and sellers' return on investment. Furthermore, in case of an accident, the sensors can help provide information on what to do and what to avoid next time.

Hazard Analysis and Critical Control Points (HACCP) is a a scientific and systematic preventive approach to food safety, from biological, chemical and physical hazards in production processes that can cause the finished product to be unsafe

Hazard Analysis and Critical Control Points (HACCP) is a preventative system for food safety that works by seven principles, identifying hazards, identifying critical control points, establishing critical limits, monitoring critical control points, taking corrective actions, verification, and keeping records. This thorough process, in line with the policies of the Food Safety Management System (FSMS), enables a sound approach to transport food from manufacturer to consumer.

Companies work under the pressure of regulations and policies such as the ISO22000 and ISO 22005 to provide perfect products to buyers. Their job is made easy through the use of ISNs, and illnesses are avoided on the consumers' part as well, thus benefitting everyone. ISNs make crop-growing more efficient while also making sure that it reaches people in the safest way possible. From production to transportation to consumption, sensors ensure quality and health.

3.4 ISNs in food warehouses and stores

Before and after food is transported to specific locations, it is stored in warehouses while the next few steps are being devised and prepared. Middlemen, traders, and producers all use warehouses to store harvested and packaged products before moving them to the market. Losses are often faced at this stage due to incompatible conditions. Traditional storage methods require a lot of manual labor; their replacement with IoT via sensors and data storage and prediction has made the process a lot easier.

IoT-based warehouse monitoring systems such as Raspberry pi[2] , which stores the data it senses on a cloud called ThingSpeak and maintains database

---

[2] Bhandari, S., Gangola, P., & Verma, S. IoT BASED FOOD MONITORING SYSTEM IN WAREHOUSES. International Research Journal of Engineering and Technology (IRJET). https://www.irjet.net/archives/V7/i4/IRJET-V7I41124.pdf. (2020, April).

Mysql are good examples. The sensors used are the DHT 11 sensor, LDR sensor, MQ 3 sensor for humidity and temperature, detection of light radiation, and that of nitrogenous gases, respectively.

IoT devices need to be installed in food stores, warehouses, etc. Once they are connected with the Internet, they start reading data from the interfaced sensors, and the job of safely storing food in a lesser time frame is made possible. The environmental conditions in warehouses are thus monitored to prevent decaying and rotting of food items, mainly wheat, rice, and maize, that are prone to spoilage quicker. Some systems are supported by buzzers as an alarm system that activates as soon as the threshold value of the sensor crosses a specific value in order to alert the personnel. Timely action is then taken to prevent any damage to the food.

Another example of a monitoring IoT system for warehouses and cold storage is Efento Cloud[3] , which is versatile enough for monitoring environmental conditions at thousands of points. As the name suggests, it also allows for storing that data for future references. It also provides a range of sensors, including one for measuring barometric pressure.

With regards to ISN-enabled stores, the example of Amazon is befitting. Amazon's purchase of whole foods was announced in June of 2017; its biggest acquisition ever to buy the Whole Foods supermarket chain for $13.7 billion. The Everything Store promises to redefine what a trip to the grocery store looks like. The primary purpose of such an acquisition is to address the need for a large, growing grocery market with the lowest friction point for customers. The secondary purpose of this acquisition is to connect the digital world to the real store/real world by which Amazon can also sell other goods at those locations.

---

[3] WAREHOUSES AND COLD STORESCOMPREHENSIVE MONITORING OF TEMPERATURE AND HUMIDITY. Monitoring Temperature and Humidity. https://www.efento.gr/en/application/food-industry/monitoring-temperature-and-humidity-in-food-warehouses-and-cold-stores/ (2021).

Amazon Go is the fast-tracked grocery shopping experience dependent on sensors and machines.

- Amazon Go

It is a supermarket with no cashiers. "No lines, no checkout," states its slogan. Embedded ISNs and other technology keep track of what you bought. So far, only Amazon employees can shop at one Amazon Go location by the company's headquarters; technical issues have kept the company from rolling it out[4] . A lot of the stuff you buy in a grocery store spoils quickly, which means you must get them home quickly—plus, someone has to be there to receive the goods. Ultimately you can monitor the freshness of the grocery to make the delivery data-driven.

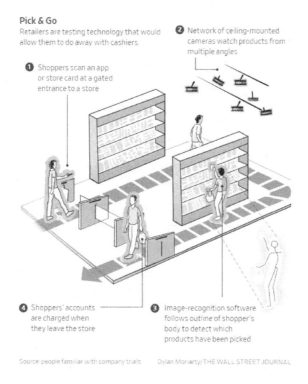

Pick & Go
Retailers are testing technology that would allow them to do away with cashiers.

❶ Shoppers scan an app or store card at a gated entrance to a store

❷ Network of ceiling-mounted cameras watch products from multiple angles

❹ Shoppers' accounts are charged when they leave the store

❸ Image-recognition software follows outline of shopper's body to detect which products have been picked

Source people familiar with company trials   Dylan Moriarty/THE WALL STREET JOURNAL

---

[4] Laura Stevens, Amazon Delays Opening of Cashierless Store to Work Out Kinks, Wall Street Journal, March 27, 2017.

Figures 3.10: Pick-and-Go-system
Source: The Wall Street Journal

Amazon Go also plans to monitor the buying patterns of customers to have an intelligent availability plan, but there were problems monitoring more than twenty people at a time in the store. Sensors in such stores would identify customers and track their movement from aisle to aisle, understanding their shopping patterns and priorities. More sensors for charging customers' mobile phones for payments would also be installed. Such utilization of this technology would change the face of the retail industry and make buying a process much quicker. To conclude, we covered the smart sensor technology in the production of food and crops, in transportation and storage of that food, while pointing out the merit and advantages it has over manual maintenance. ISNs make monitoring of all kinds more manageable, efficient, and less time-consuming as compared to the traditional ways.

3.5 Putting it All Together

With the various examples that we have talked about, we can safely assume that with the development of wireless sensor technologies in food markets, food production, food storage, the overall increase of need and interest for IoT systems in the population is growing fast. And, since high-quality food, health benefits, safety measures, and the growing demand for healthy and "clean" products are on an all-time high, the use of sensors is bringing usage to a new level. Moving away from Conventional equipment, the Internet-of-Things (IoT) has helped add a level of technology to the Food Industry. A broad series of networked sensors, monitors, and other Internet-connected devices, IoT technology, are being used to oversee the entire food manufacturing and distribution process from the warehouse to the point of sale. With sensors detecting the shelf life of our food to the food manufacturing sector, using digitally connecting supply chain systems to provide greater visibility into the production of foods and beverages, everything can be used to improve the overall quality of the food that people consume.

Putting it all together, when we look at the future of how food is processed, kept, and stored, we see a massive change from the times before. In the future, IoT-based technology will likely impact virtually every aspect of the food sector in some way. However, we also see that the application of new technology or technology as a whole in food is more sensitive than in many other industries. People that are a part of this industry see many possible outcomes and possibilities to make it even better, but they are also aware of the hurdles they have in their way. However, with strong research, detail-oriented work, and ever-improving IoT-driven automation, with time, the food industry and its revolutionary use of the ISNs system will definitely help consumers make better choices about the foods they consume and purchase.

And as these technologies continue to become more intelligent, the food industry's future can be summarized in the example picture shown below.

The equipment and components reviewed here were specifically chosen because they are the most used by the organizations of this industry. Food processing, specifically IoT-based applications that use technology, science, and real-time data streams to make the whole process better, offer multiple benefits, including:

- Improved Food Safety: More IoT based technology equates to less human interference in the production process, which reduces the risk of contamination

- Thorough farming land check: An over-head drone or aircraft provides plant protection by comparing real-time data and assessing the requirement by the soil it's covering.

- Services like SEE™ Advanced Maintenance Program: They help identify inefficiencies within the production line and provide actionable solutions and technical expertise.

- IoT-based warehouse monitoring systems: Such offers various solutions for monitoring and safekeeping of food items and take instant remedial actions if necessary.

- Better transportation: With cold-chain, food processing preservation in infrastructure facilities has improved. The cold chain provides safety while transporting food from the starting point. It covers the creation of infrastructure facilities along the entire supply chain until it reaches the consumer.

Chapter 4: Intelligent Homes, Offices, Roads, and Buildings

"The ache for home lives in all of us, the safe place where we can go as we are and not be questioned."

Maya Angelou, American poet

"We get insights from the fridge. It has a friggin' camera in it. ... Wow, this guy really loves onions. He's Shrek! Let's give him more things with onions!"

Lisa Fetterman, founder/CEO, Nomiku

4.1 Background and Introduction

An intelligent home/building or a smart home/building is any residential, commercial, or agricultural setup where the building allows its inhabitants to leverage technology to enhance their lifestyle and improve building functionality. It uses remote, internet-connected devices to provide you greater independence while maintaining good health and preventing social isolation. In an aging world, where technology is slowly taking over everything, the concept of an intelligent home does not sound inane.

In fact, to the contrary, from the concept to its implementation, intelligent homes and any related living structure sound essential. If we look at history, infrastructure and the setup of homes were not always smart. When appliances were first invented, they went through an extensive stretch to provide comfort to their users. However, since the advent of the 21st century, smart homes, smart buildings, roads, and even regular home automation began to increase in popularity. And, from the early 2000's we can see different innovations beginning to surface in the world of organization and homes. From thereon, Smart homes suddenly became a more affordable option, and therefore a viable technology for consumers. From domestic technologies to home networking to other gadgets, technology began to appear on store shelves for even

77

everyday household users to consume. And, now smart cities have started to emerge in every corner of the world that uses the Internet of Networks (IONs) and ISNs to plan big urban data-induced projects. One of the countries that already use ISNs is Japan, where several projects currently aim to maximize the use of assistive technology, enabling the elderly to live autonomously at home by creating a smart and comfortable environment.

In the following sections, we will review how these multiple ISNs work to provide a better life for their users. From homes, large office buildings to the roads we travel on, we will assess just how these modern-day sensors, their working, quality, and innovation will close the gaps between the current weak points of ISNs and the predictive future of this technology.[5]

4.2. Home ISNs

The concept of using sensors and converting your home into a more intelligent home is a promising and cost-effective way of improving your everyday life. Intelligent homes are developed in a way that they are equipped with sensors, actuators, and biomedical monitors if required. It can range from any technology you might require to any technology that's available in the market right now to improve your living standard. Earlier, the concept was used for anything that used technology. The most famous and in-use home automation concepts of today include remote mobile control, automated lights, automated thermostat adjustment, scheduling appliances, networked appliances, notifications on your mobile devices, and remote video surveillance. It's safe to assume that most users of smart homes use ISNs for security purposes to adjust their temperature with just a click of a button. But now, with changing times, the concept of smart homes extends well beyond simple automation or appliance control.

[5] H. Kelly. "The scattered, futuristic world of home automation", CNN Business . 2013.https://edition.cnn.com/2013/01/12/tech/innovation/future-home-automation/index.html

ISNs will help the users to manage, foster the safety, security, comfort, and health of the residents. The goal of the new smart systems is to provide state-of-the-art connectivity to every household, even when you're not home, and in the future, we can look forward to cutting-edge technology not dissimilar to the one seen in the animated series "The Jetsons." They say that the technology of then is finally a reality, with major vendors peddling real, usable products, almost all controllable from a smartphone. We can also look forward to digital cutting boards (digital everything, really), molecular cooking devices, and so much more.

But how do these devices work? Their mechanism is fairly simple. For any home to be intelligent, wireless sensors are attached due to their several obvious advantages. These sensory devices that are connected to create smarter, more intelligent homes do need a system to function properly, but they don't need any formal training or intricate education to be worked with. These devices operate in a network connected to a remote center for data collection and processing. But, unlike Wi-Fi, they are not limited to just a specific area range. With technology advancing and better equipment becoming available on the market, they can be used to create a network of devices that communicate with each other, pass on signals from objects to your device even when they are in far-off areas of the home. A few of the advantages of these wireless sensors are that they can be easily installed, maintained, and connected inflexible, reconfigurable networks, and they can be adapted to the users' needs.

Your smart home needs vary depending on the occupancy. For any purpose, the remote center diagnoses the ongoing activities and initiates assistance procedures as the users require. The technology can be extended to wearable and in vivo implantable devices to monitor people 24 hours a day, both inside and outside the house. In today's world, when safety is a man's biggest priority and need, ISNs can also help alert us to any outside intruders.

However, we also have to realize that this trend of individual devices and sensors having their own smartphone apps and cloud infrastructure sounds great right now when we only have to keep track of a handful of devices, but in the long run, when everything will become digital, this idea is not just unsustainable, but it's also a little unrealistic. Because not many people have more than a dozen apps on their phone and even if they do, imagine having to sort through thirty or so apps to set your thermostat to the right temperature or find the right lock, or opening each app one at a time to set up any kind of home settings when you're going out for vacation. [6]

4.2.1 Smart Home Devices

Today, as smart technology continues to improve our living standards, many smart home devices have become available in the market. In other words, home automation has been ingrained into our lifestyles. The trend shows no sign of slowing down— according to IDC, as many as 832.7 million smart home devices have shipped in 2019. By 2023, that number will climb to a staggering 1.6 billion. However, as the requirements and preferences of consumers change and people gain an understanding of sensors, it has become more of a need than a luxury for them to upgrade the technology of their homes to provide them an overall efficient environment to live in. The good news is most of them are available on the internet for an average do-it-yourselfer to install, buy and use safely. Some of the most common easy-to-use categories include:

Smart Heating and Cooling System

As the name implies, this technology automates your home's heating and cooling system for optimal comfort while helping you save money and energy. And, traditionally, that's all it did – heat and cool our indoor spaces unevenly, but now with better HVAC systems (Heating, Ventilation, and Air

---

[6] H. Ghayvat et al. "WSN- and IOT-Based Smart Homes and Their Extension to Smart Buildings." Sensors (Basel, Switzerland) vol. 15,5 10350-79. 4 May. 2015, doi:10.3390/s150510350

Conditioning), our indoor environments are becoming more comfortable, controlled by smaller zones, and more efficient to maintain. These HVAC systems are composed of many subsystems; each of them may exhibit nonlinear feedback characteristics, which means it does not necessarily wait for instructions from you in order to make a decision. Instead, it makes sudden changes keeping in mind the development of its surroundings. The parameters of the systems change with the outside weather condition, inside load, and process occupancy. Most HVAC systems are controlled by four input parameters: humidifier pump controller for humidification, fresh and bypass damper controllers for temperature control, exhaust fan controllers for the stability of differential air pressure, and ventilation fan controllers for airflow capacity. This, combined with weather systems for the local region, can become a more efficient and better controlled closed-loop system.

Smart Locks and Home Security Systems

A smart home system must keep your safety as the main priority. In this day and age, it's absolutely essential to upgrade your home safety. The existing security model sounds simple, but even within that are incorporated many complexities that work to best serve you and ensure your ultimate safety. Smart door locks are one of the most critical pieces of equipment adopted to enhance home security. These door locks use battery-powered sensors, monitors, and many today also have the feature of biometric recognition to digitally unlock your door. All of these features are embedded to secure and protect your home from any unauthorized entry. Once installed, they send real-time alerts every time someone enters your home.

Home security systems are not just restricted to using door locks. In reality, something as basic as a security camera that has been around for years is also one of the most utilized home automation equipment since cost, performance, resolution, and interface are being upgraded every three months. You can now buy a 4K HD or 8K HD security camera with built-in face recognition and motion detection. Motion detecting technology has upgraded a traditional security camera to serve the user better and identify exactly what's happening in real-time by notifying you when it detects any motion. Rather than recording for 24 hours and storing all that extra video footage, a motion detection camera is activated by a motion sensor. Motion sensors typically use PIR detection— passive infrared motion sensor technology.

A PIR motion sensor is the least expensive type of motion sensor, which works by looking for any infrared radiation that comes close to its sensing radius. The motion sensor can detect infrared radiation, which is a naturally emitted radiation exuded by all living creatures in the form of body heat. This feature allows motion detection cameras to safely differentiate between a tree branch and a human being that might be standing in front of the camera. And, if there is any radiation detected, this radiation is picked up by the sensor and triggers the safety alarm.

Smart Plugs

A smart plug at your home monitors energy usage for outlets. It can be plugged without any hard work into any regular wall power outlet, and once it's connected, you can control every device that you later plug from your phone or laptop. This concept of the smart plug dates to 2008, where Dr. John Woods, one of the researchers working on smart plugs and their technology at the University of Essex in Colchester, presented an 'intelligent' plug for energy savings and energy efficiency at the university's 'iSpace' department. It works like a power-point adapter that lets the central hub, which can be your phone or any other device, control its every function. Such functions can range from increasing the television volume from your phone to turning off your lamp, your straightener, or just about any appliance connected to the plug without having to move. Smart plugs are now Wi-Fi friendly, and some can also be compatible with your virtual assistants like Alexa, Google Assistant, and Samsung SmartThings.

Smart Speakers and Displays

Smart speakers paved the way for virtual assistants and displays to become part of our lives. Now, owing to their features and usability, every big company have their own version of it. These smart speakers and virtual assistants have quickly become one of the most found appliances and the center of every smart home and office. Speech, also known as voice recognition technology, has come a long way since its inception more than 70 years ago.

The key technology behind the smart speaker is voice recognition. They are ultra-sensitive and always in "on" mode and ready to listen to you using an array of microphones and noise-canceling technologies. As you might have Seen, Microsoft announced the acquisition of Nuance in April of 2021, which is one of the most widely speech recognition software in the industry. As these smart speakers sense the user's voice to convert the data from analog to digital, then it is sent to the cloud for interpretation or recognition of words and sentences. It encrypts the structure of the user's command and content using Natural Language Processing(NLP), a procedure of converting speech into words, sounds, and ideas to convert those words into meaning. The sensors will also analyze the characteristics of the user's speech, such as frequency and pitch, to give you feature values. Then, it takes the context input and decides the response based on the command they receive from the cloud and responds to the user by using Natural Language Generation technology. These virtual assistants work on Natural Language Technology techniques in which they try to remember what they have heard or done in the past in order to improve their response to future requests.

However, virtual assistant technology can also be used to communicate with other cloud-connected smart devices like your television by using an Application Programming Interface, a software intermediary that allows two applications to talk to each other. With the help of a smart device that supports an AVI, you can control many of your devices through routines.

4.3 Industrial Building ISNs

The Internet of Things is constantly making the devices that we use more intelligent and more connected. However, this automation technology is slowly taking over every industry. And a popular area where IoT systems are being set up are Smart buildings. With every passing day, they are becoming a hotbed of innovation and attention. It's one of the areas where we can see the impact worldwide.

The basis of differentiating a smart building from a regular one is fairly simple. Smart buildings are categorized using wholly integrated systems which means it brings together the component of subsystems into one system, that share vital information. These Smart Buildings use IoT sensors and technology, a floating floor or raised floors, and your everyday home ISN's automation to control nearly every building component. They can control your HVAC, lighting, shading, kitchen appliances, gym equipment, reception duties, check-ins and check-outs, security (which for many is the biggest upside of a building-wide ISNs), and even user-centric functions like wayfinding and conference room scheduling.

These buildings' control and automation are done primarily to assure efficiency, energy use, ease of use, comfort, access, and monitoring for the businesses or residential hubs that dwell in them. A BAS, also known as Building Automation System, is one system or set of systems that provide automated control and monitoring within a building, and it relies on network infrastructure elements like;

- Wired and wireless networks to control the data and information in real-time.

- Licensed and unlicensed spectrum and software to manage them.

These features are the most basic but also the essential building blocks of seamlessly integrated fiber, copper, and high-speed wireless networks that combine to form the central nervous system of the entire building automation system to deliver crucial information.

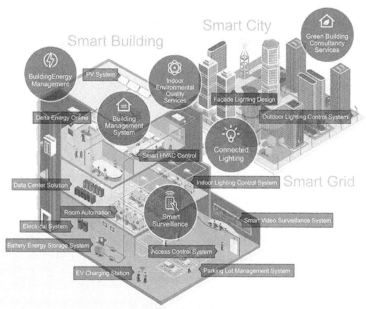

Figure 4.1- Showing the context of smart building and smart city- Source: Delta

As shown in the figure 4.1 infographic diagram, many offices have some level of building intelligence. They can be simple sensor-based check-in and out systems for their employees, lighting and heating systems, and some even more advanced complex technologies making everything use IoT for smarter surveillance and an overall better experience. In a simplistic model, this may look like most buildings in the world use this model, but they don't since Industrial buildings have a wide range of monitoring, management, and resource optimization requirements. In order to be operative and in effect, the goal is for any automation system to enable all these ISNs systems to work from a single building control point and build a complete smart industrial hub.

Technological conjunction as it relates to industrial building management and the connection of smart buildings is accelerating with the increasing deployment of IP-based devices under the thrust of IoT. A few years ago, various building systems utilized different protocols, networks, and cabling systems. However, now these systems can be seen worldwide. Whatever these technologies may be, it's no secret that the buildings of today are surely changing, and a few buildings already do so much more, and they're getting smarter. Some of the examples of these automated buildings are:

Capital Tower, Singapore.

The Capital Tower has several intelligent ISNs enabled application that makes it a smart industrial building. And from a green, eco-friendly perspective, it may be the ideal building. Starting from energy efficiency systems that are built-in, it also includes an energy recovery wheel system in its air-conditioning unit, which allows cool air to be recovered to maintain the chillers' efficiency. Motion detectors installed at the lift lobby and toilets conserve energy, while double-glazed glass windows reduce heat penetration and minimize energy consumption.

The Edge Building (Amsterdam, Holland).

This building employs state-of-the-art digital sensor technology to measure and keep track of occupancy, movement, lighting levels, humidity, and temperature, and using smart technology – including Ethernet-powered LED connected lighting to keep the light level of the building exactly how the people inside the building requirements. To understand its model, we can say that the building systems respond to maximize efficiency. The IoT in the buildings plays a role in facilitating the injection of smart "things" in the environment of the organization.

In simpler words, this building knows more about you than your co-workers. It knows where you live; it knows what model of the car you drive, what time your car enters and exits the premises so it can predict future patents. The model of the building will direct you to the most convenient parking spot;

it has enough mind of its own to know your schedule for the day and even how much sugar you like in your coffee.

Hindmarsh Shire Council Corporate Centre, Australia.

In designing Hindmarsh Shire Council offices' smart building, Melbourne-based architect firm k20 Architecture wanted to improve energy efficiency while also enhancing the office environment for employees. The building is located in Melbourne, an area exposed to extreme temperature conditions and intense sunlight. The architects at k20 Architecture - Architizer wanted to use this temperature "disadvantage" to their advantage. The obvious answer was to install solar panels to harvest energy from the sun, but they wanted to do more than that.

In order to make a smarter, more efficient building, they made use of both technology and nature. They built a series of underground thermal chambers, where the environment is controlled, and then installed a ventilation system under the floor to help draw fresh air from the outside. The earth naturally cools or warms the air according to the temperature, after which their air is redistributed back through the building. In addition, LED lighting systems reduced energy consumption and maintenance by using sensors to turn off any lights where human presence was not sensed. Crossflow ventilation and zoned motion-detecting lighting also improved energy efficiency, while vertical green walls enhanced indoor air quality.

4.4 Workplace ISNs

In the workplace, as devices continue to connect employees, the IoT is changing the way how employees interact and connect. Workplace ISNs are creating an environment that is much more than a simple four-walled office – these IoT's are creating collective intelligence in creating a better workplace.

Applications abound in workplace safety, employee productivity, asset management, and building comfort. By thinking strategically about how to "intelligentize" the workplace, companies are leveraging their greatest assets – their employees – to work smarter and be more productive.

One of the ways that a workplace is getting more intelligent is to see their use of sensors and analytics. For example, an employee's work behavior while he/she is connected to the server can give real-time information about his/her timing of productivity. It can help the organization to make decisions regarding what time to schedule assignments and important client meetings depending on the employee's productivity hours.

These possibilities are exciting, but just like everywhere, even in the workplace, the use of IoT presents several challenges.

For instance, the proliferation of earliest-to-market IoT devices and services has led to interoperability issues, which means some organizations struggle with devices that can't communicate, data that can't be shared, and updates that are inconsistent [5]. Meanwhile, security is always a risk factor when it comes to the IoT, as they rush to be the first has led many device makers to treat security as a secondary thought. Currently, when everything had been digitized, it still does not make us any less worried about our sense of security. The same is true for employees. In an IoT network, tracking and monitoring employees, even for work purposes or calculating their workplace efficiency, can involve privacy and data concerns that most businesses have not faced before. [7]

4.5 Smart Cities

The smart city idea coordinates data, figures, and correspondence innovation, Information and Communications Technology (ICT), and different available gadgets associated with the IoT organization to upgrade city tasks and administrations' productivity and interface with residents. Shrewd city innovation permits city authorities to collaborate straightforwardly with both

---

[7] Ron Exler, IoT and the Intelligent Workplace

local area and city foundation and screen what's going on in the city and how it is developing. Effectively, an organization of Internet of Things (IoT) gadgets and sensors associated with server farms produces information that cultivates examination, knowledge, and experiences for savvy city applications and administrations. According to data, the world will associate an expected 24.1 billion IoT gadgets worldwide by 2030.

IoT-empowered applications and administrations, for example, associated travel and bicycle shares, shrewd stopping, associated CCTV, environment-specific lighting, and brilliant locks, make sensitive urban areas, grounds, and structures more secure and more proficient to oversee. Furthermore, urban communities can screen air, water, and contamination quality to improve general well-being. At the same time, keen home and keen office encounters help expand property estimations and rents while diminishing operational expenses. One of the activities ready to shape savvy urban areas is the slick light shaft that serves as a streetlamp and is incorporated with security, organizing, force, interchanges, and IoT advancements given by different specialist organizations.

4.6 Hospital ISNs

With every business sector worldwide going toward digitalization and adapting technology, healthcare delivery is no exception. And, since due to it, the whole paradigm of care delivery is changing. With the increasing frequency of medical data gathering and documentation, every passing day, an enhancement in healthcare services is required to fill arising gaps or errors. The patients need these hospitals to work at greater efficiency, better overall care, and comfort, and this need is reshaping healthcare systems globally and encouraging this digital transition. And it is why the concept of smart hospitals is not a new one.

The smart hospital is the future of care and provision. When it comes to the supply side, there is a host of new technologies that can now be integrated into care delivery to meet the health needs of the population. Some of these hosts of technologies include artificial intelligence (AI), robotics, precision medicine (an approach to patient care that allows doctors to select treatments that are most likely to help patients based on a genetic understanding of their disease), 3-D printing, augmented reality/virtual reality, genomics, telemedicine, and more.

Adoption of these technologies is being driven by immediate needs (e.g., cost control and efficiency optimization) and longer-term goals (especially greater precision, fewer errors, and better outcomes).

Sensors and these new-age interaction technologies flawlessly integrate into such environments.

These advances, especially in the medical field, offer various forms of personalized and context-adapted medical support, including helping patients on procedures that are not risky. By smart hospitals, we do not only mean that the equipment these hospitals use is made with sensors, rather than the entire environment of the hospital from the reception to the operation theatre works to increase efficiency, improve the care and make the process of going to a hospital much easier. It lends you the assistance to carry out everyday activities, for example, monitoring personal health conditions, medical and emergency systems, enhancing overall patient safety, as well as getting access to social. By providing a wide variety of services, smart healthcare applications bear the potential of bringing medical, social, and economic benefits to different stakeholders. The goals are from enhancing comfort, supporting autonomy enhancement up to emergency assistance, including detection, prevention, and prediction.

Smart hospitals do not attempt to be the one-stop solution. They don't deliver all services under one roof. Rather, they deliver a narrower set of high-value services within a broader ecosystem of entities, many of which have not traditionally been associated with healthcare delivery until today. In such an ecosystem, preventive services and healthcare management programs are delivered straight to the patient. For instance, these services can be delivered at clinics, gyms, and even in patient's homes.

Other medical treatments and minor procedures are provided at ambulatory centers. Diagnostic testing (imaging and laboratory services) is offered at independent centers. Hospitals are responsible only for major surgeries, intensive care, the management of severe trauma, and treatment for other acute, severe, complicated conditions. According to health care global, by the year 2030, the world will be home to more than 9 billion people, so as the population of the world increases, the pressures on the healthcare system will eventually increase as time goes on. This population increase, therefore, contributes to the constantly rising costs of health care, and this is where a smart hospital comes in to reduce these costs.

4.7 Airport ISNs

Air travel has grown considerably in current years and is expected to keep growing in the medium term. Billions of travelers skip through airports every 12 months. However, with the crisis of the pandemic raging, the world sees a need to switch to a more technologically friendly airport. From the need to restrict crowds in airports, the current logistics and infrastructure models of airports can use linked technologies to extra efficaciously put into effect effective passenger, baggage, and plane processing structures to avoid congestion, spread the demand through the years, and better assess fitness risks, while at the same time optimizing processing times. Studies have shown that a person wastes an average of 16 days waiting in airports in his/her lifetime (6). This is time that you won't get back, and this is time that we can save by adopting technology in air travel.

To make the passenger journey a lot easier, the concept of smart airports has been introduced.

Smart airports employing IoT promises to have an enormous positive impact economically, environmentally, and many other aspects. Despite this, though extensive research and work have been put into sensors and IoT-based smart cities, practically no work has been done on the realization of a complete smart airport system to track passengers, luggage, parking, map, locations, and other facilities.

Internet of Things (IoT) devices and sensor technology are trying to make air travel more efficient. This can enable better accuracy in maintenance work as well as better managed maintenance, and thus the safety of the plane can be improved. The IoT devices collect data on the plane's current conditions, helping provide a safer environment for the travelers as well as the workers since in case any fault occurs, it can be handled immediately. IoT-based smart airport solution enables controlling and monitoring many systems from a remote area, unlike in conventional airports.

Some of the proposed benefits that a smart airport can give to its users are as follows:

- IoT will save time; it will enable travelers to know which lines are the shortest, helping them not miss their flights.

- It can help them immediately find a parking space in the airport and self-checking luggage.

- The lengthy-term development of smart airports may encompass the use of added self-reliant solutions, hence increasing and optimizing airport hours of operations and improving overall protection.

- Activities including delivering luggage, fuelling planes, site visitor's control, immigration and security methods, and clearing debris will stay completed with growing autonomy.

4.8 Putting it all together

"Anything that can be connected will be connected."

This quote essentially explains the path the world of today emphasizes. As we aim to connect the world, we understand that our intentions to make a smarter world deliberately rely on the concept of the Internet of Things (IoT). IoT holds the potential for us to effectively help build a "smarter" future. IoT-based online monitoring approach using smart logistics can address a lot of the crises that arise in the infrastructure development of our cities. It is no secret that the use of IoT to make the world around us smarter has a lot of advantages:

- Improving efficiency and reducing the cost of living.

- Enhancing user experience.

- Improving the monitoring of security and safety.

- Improving operational efficiency by using data on the current living pattern, no matter where you live.

However, it is not to say that the road to achieving a smarter world is easy. There are several challenges to making this technology work reliably in the highly dense and dynamic environment of real-world logistics operations. Since a cloud service will be used as the main database for the system to be controlled, it brings a risk of safety and protection, which can make the user skeptical.

Chapter 5: ISNs and Our Mobility

" When you care about people's happiness and productivity, you give them what brings out the best in them and their creativity. And if you give them a choice, they'll say, 'I want an iPhone,' or 'I want a Mac.' We think we can win a lot of corporate decisions at that level. "

Tim Cook

June 7ᵗʰ 2010--- "Steve Jobs just announced that the new iPhone4 will include a gyroscope, which should be a great addition for gaming, augmented reality, and even 3D apps. Apparently, the iPhone 4 beat this industry insider's prediction of gyroscopes in a mass-market phone by 2 years. "

The Next Web

5.1 Background and Introduction

In previous encounters, we have studied how ISNs help users in different fields of life, whether it's the food industry, the housing industry, or the health care sector, and we have gone through the uncountable applications it has had in these fields. However, with the emergence of portable devices like mobile phones, smartwatches, and electric cars, this fast-pacing information age has wrought one of the deepest social revolutions of the 21st century, where we see ISNs systems attached everywhere. The IoE, with the help of wireless sensor networks, allows different electronic devices with different capabilities to sense the environment and to communicate for data exchange, and millions if not billions of users are engaged in these devices every day, all over the world sharing their data that these ISNs systems then store,

In the case of our movements, the main content of these ISNs is the so-called mobility data that they collect based on our movement, i.e., information and any data about the movement of these objects. The least that information can tell us is our location and time information.

In this portion, we will preview the concept of mobility data, mobile devices and later see what we can learn from such data collections. These mobile devices may have different classes, types, and capabilities, but the data and information they store are what the user should know about. So, before we talk about the different kinds of mobile devices and the ISNs systems they have, we need to know what mobility data really is;

Mobility data in its simplest form could be a variety of information:

- Data coming from our mobile phone conversations (my provider knows that I'm located somewhere during the period of my call).

- Data recorded by GPS devices during an activity (my device records that I'm at a specific location at a specific timestamp).

- Data exchanged between vehicles in a vehicular ad hoc network (VANET) environment, which is basically a group of moving or stationary vehicles connected by a wireless network (a vehicle 'asks' its nearby vehicles ask information regarding the route, e.g., the cheapest gas station in the neighborhood).

- Data collected by Radio Frequency Identification (RFID) systems. They use sensors that are supported by electromagnetic fields to automatically identify, and track tags attached to objects, so when an RFID-equipped parcel passes an RFID reader, the information that this object passed from that location at that timestamp is recorded in a database.

- Data extracted by smartwatches and smart clothing with the help of Wi-Fi access points; the Internet access provider knows that as long as you are served by that access point, you are located at a specific location.

Mobile devices combine our need to move with a variety of sensors that provide them with the necessary framework, data, and any information these devices can acquire about us for a more comprehensive approach to remote monitoring. And, by understanding how mobility data and the sensors attached to these devices work, we will look at how these devices function by discussing the transition from— stationary—to mobility data management and the challenges that emerge with this new form of technology.

5.2 Mobile Phones ISNs

There is no one device that is more important and more popular than our mobile phones, and especially in this season of the pandemic when meeting other people and our daily face-to-face interaction has been restricted to a minimum, phones have been our savior. They have given us the technology, the science, and the means to stay connected with the rest of the world. And, as technology continues to evolve, more and more physical devices are being integrated with sensors and connectivity. However, some of the widely used devices that work primarily with the help of sensors are mobile phones. According to the latest report from GSMA Intelligence, there are 5.22 billion unique mobile phone users in the world today, which means there are more than 5 billion people in this world that are in touch with ISNs and their technology in one way or another. These mobile phones that we use are equipped with multi-level communication interfaces, such as Wi-Fi, Bluetooth, near-field communication (NFC), and cellular communication that help us make calls. To understand a mobile device and its ISNs, we need to understand that every mobile phone has several sensors to monitor different features and to detect change. These sensors consist of precision components, which are sensitive to sources of external interference and physical factors. These uses of sensors are the same in different mobile phones, no matter who the manufacturer is.

Sensors in a mobile phone work just like they do in most devices we have talked about before, but an overall concept of sensors is to enhance smartphones' usability. However, when it comes to mobile phones, their sensors and tracking systems are much more complex.

Common mobile phone sensors include:

Fingerprint sensor

Function: Verifies your fingerprint for screen unlock. It is one of the widely used sensors in new-age mobile phones.

How it works: These scanners use optical fingerprint data, combined with Capacitive fingerprint sensors, commonly used on devices to detect a real finger. These sensors sense the electrical current produced by finger touches and then generate an image of the ridges and valleys that make up a fingerprint. The sensor then compares the image with the stored version on the device, which means it goes through the history of all the fingerprints it has in its memory, and if one of them matches, it unlocks the device. In-screen fingerprint sensors, which are also known as optical fingerprint sensors, sense the fingerprints using light reflections. They are used with OLED screens where the spacing between pixels of an OLED allows for the transmission of light. Every time a user touches the fingerprint sensor icon, the OLED lights the touched area and starts the 'scanning processes.' Then, the in-screen sensor underneath the screen grabs an image of the fingerprint that was projected onto the sensor. The sensor then does the same thing a capacitive fingerprint sensor does and compares the image with the stored version on the device.

Linear sensor (accelerometer)

Function: Allows your device to switch between landscape and portrait modes automatically, helps count your daily steps, and identifies your viewing orientation, so if you're laying down, your phone changes its display mode in accordance with the new angle. This sensor is used to determine the device's relative orientation in space, so it can be used with compass apps and recognize motion gestures (such as picking up and flipping your device).

How it works: Most devices use a gravity sensor now by measuring changes in the distance between the capacitance plates caused by motion on three axes (X, Y, and Z) and determines the instantaneous acceleration and deceleration forces accordingly.

Ambient light sensor

Function: It is a photodetector that automatically adjusts the screen brightness of your device according to the amount of light present in your surroundings, making it more comfortable to look at the screen. Each device has its own specific operating range of performance, from very low light up to bright sunlight but, in good quality phones, the sensor not only supports automatic white balance (AWB) when you are taking photos but also works with the proximity sensor to prevent your phone's brightness from intensifying when, for example, the device is in your pocket, and the ambient lighting is close to nonexistent.

How it works: The sensor generates strong or weak currents according to the amount of ambient light it senses, and the device increases or decreases the screen brightness accordingly. Please note that using a non-official protective case or unlatching protective film may block the ambient light sensor and affect its functions.

Proximity sensor

Function: A phone with a proximity sensor detects the presence of nearby objects. Any device equipped with a proximity sensor automatically turns off the screen when it detects that it is close to your ear. This helps prevent possible mis operations.

How it works: The sensor consists of an infrared LED light and an infrared radiation (IR) detector, and it's generally located at the top of the screen and near the receiver so it can detect human presence. The sensor detects the distance between an object and the device by calculating changes in the infrared light signals it receives; the closer the object comes, the more changes there are in the IR, and the sensor picks up on these changes. The working range of a proximity sensor is generally 10 cm.

5.2.1 Sensors in Mobile Phones and the Security Predicament

In the era of IoE, data content is stored not only in the device but also in Internet servers which opens the door to a lot of security threats and mishaps. Since these applications store their data on the cloud, they require mining of hidden information of the smartphones and their sensor data. In other words, the data sensors collect through the user's pattern and behavior, which are then leveraged in new indirect ways to predict, help and estimate new features that are not directly designed to be assessed by these sensors. This means that this data is being accessed by more than just our device. This new usage pattern of smartphone sensors exposes privacy and security issues since smartphones are multiplying by the minute. The users are getting more and more concerned about their safety and privacy because there is an evident demand for tougher security.

5.3 Automobile and Truck ISNs

Mobility and this freedom to move wherever we want whenever we want is not just vital, it is the backbone of urban life and a crucial economic factor in our world's development. And now, in this fast-paced, technologically driven world, transportation is another sector that has become the hub of lasting change. It is a sector as old as human life but with rapid urbanization, the progress of mega-cities, and the conception of smart cities. All this evolution is bringing dramatic changes in the capabilities of vehicles and how they operate in the world. Innovative solutions like autonomy in the shape of self-driving cars, electrification, and connectivity are on the rise, and this desire that your car will be another medium of sharing real-time data on the cloud.

The automotive sensor markets

Automotive sensors are used to detect and monitor different physical and chemical processes in a vehicle, which helps find out any issues that could occur in the future. With the help of automotive sensors, leading companies provide safety, comfort, and affordability to their consumers. According to a new report published by Allied Market Research, titled "Automotive Sensor Market by Technology and Application: Opportunity Analysis and Industry Forecast, 2019–2027", the global automotive sensor market size was $16.40 billion in 2019 and is projected to reach $37.65 billion by 2027, to register a CAGR of 10.2% during the forecast period. Asia-Pacific is expected to be the leading contributor to the global automotive sensor market, followed by North America and Europe.

5.3.1 Autonomous Vehicles Market Size

The global autonomous cars market reached a value of nearly $818.6 billion in 2019, increasing at a compound annual growth rate (CAGR) of 12.7% since 2015. The market is expected to decline from $818.6 billion in 2019 to $772.8 billion in 2020 at a rate of -5.6%. The decline is mainly due to the pandemic, lockdown, and social distancing norms imposed by various countries, which caused a dire economic slowdown across countries owing to the COVID-19 outbreak and the measure they rolled out to contain it. However, this dip in the market is only temporary, as the Autonomous **Vehicle Market** is projected to surpass USD 65.3 billion by 2027. In these numbers, the North American region is predicted to command the largest **share** in the global **Autonomous Vehicles Industry.**

Many researchers interested in the automobile industry see the growth segments for investment in this sector. As per their understanding, there is an increase in demand for intelligent sensors in vehicles. This is a market that not only has great potential but solid numbers, facts, and figures backing it because the market is then expected to recover and grow at a CAGR of 12.7% from 2021 ad reach $1,191.8 billion in 2023, and these predictive numbers allow manufacturers to grow in this automotive sensor industry[8].

---

[8]Autonomous Truck Market Size to Hit 2,013.34 Million by 2027; Rising Installation of Advanced Radars, Sensors & Cameras will Boost Growth, Says Fortune Business Insights™

5.3.2 Types of sensors and their applications.

Nowadays, the cars that we see on our roads are equipped with a lot of sensors. We don't even think about some of them anymore, such as coolant temperature, engine speed sensor, oxygen sensor, tire pressure sensor, etc. These sensors are vital for the working of a car to monitor the overall engine and vehicle condition. On the other hand, apart from these must-have sensors, many luxury cars are fitted with dozens of sensors and are connected to actuators to control the respective systems without driver intervention. E.g., ABS (Antilock Brake Systems), Traction control systems, OBD (On-Board Diagnostics), Engine management system, ESC (Electronic Stability Control), Smart sensors which are used in accelerometers, and are equipped in vehicles to provide benefits such as safety and fuel efficiency. etc. Now, with the commencement of IoT in automobiles, one can access his vehicle position and state at any moment just by using their mobile device.

But how are individual vehicles that we see on the road function, and how can we put that information to work?

Well, for starters, the process starts with these devices, in this case, automobiles that securely communicate with an Internet of Things platform. Since the urban fleet of vehicles is rapidly evolving from a collection of sensor platforms that provide information to its users and upload sifted (e.g., GPS location, road conditions, traffic, etc.) to the cloud- specifically designed for vehicles – also known as the Vehicular Cloud. Now, this information that was fed can be accessed to a network of autonomous cars that exchange their sensor inputs among each other in order to optimize a well-defined utility function. In the case of autonomous vehicles, this function is not just improving the functionality of the automobile industry but also encourages prompt delivery of the passengers to destination with maximum awareness, safety, and comfort and causes minimum strain to the human mind. In other words, what we are witnessing in the vehicle fleet is the same evolution from Sensor Web (i.e., sensors are accessible from the Internet to get their data) to the Internet of

Things (the components with embedded sensors are networked with each other and make intelligent use of the sensors).

5.3.3 Evolution of Sensors in automobiles.

The urban fleet of vehicles is evolving from a collection of sensor platforms to the Internet of Autonomous Vehicles. Applications for in-vehicle communications range from safety, location services, and comfort to entertainment and commercial services. Every day we see manufacturers coming out with cars that give drivers a higher level of safety & security, reliability, and information & entertainment. We see growth and an increase of this technology and the IoT formed by all components, large and small. We can also see evident evolution from intelligent vehicle grids to autonomous, Internet-connected vehicles all powered in the vehicular cloud, but just like in every field, there is that question of "What's Next?"

The simple answer is innovation within autonomous cars and a future where we see roads filled with self-driving cars. The truth is even though these cars are available, and we might see them every so often on the road, these autonomous cars developed by researchers and automotive pioneers are still far from commercially available for personal use when communications, storage, intelligence, and learning capabilities to anticipate the customers' intentions will all be handed over the vehicle. It's not to say that the user won't have control, but the smart vehicle will become the core system of the car.

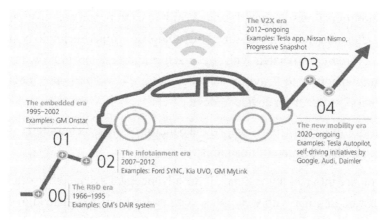

Figure 5.1: Phases of evolution of the connected car

5.4 clothing ISNs

The latest progressions in biosensor technology involve the incorporation of both wearable computing and Internet of Things (IoT) technology. It has renovated the traditional apparel industry with this development and application of smart clothing. The area of smart clothing is expected to keep on expanding in the days to come. When it comes to growth, this is predicted to occur in two distinct directions:

- performance-driven smart clothing

- Fashion-driven smart clothing.

Recently, the interest in smart and interactive textiles has grown exponentially. The global market for smart fabrics and intelligent textiles was valued at US$ 1,143.09 million in 2019 and is predicted to grow to reach US$ 6,418.08 million by the year 2027. It is estimated to grow at a CAGR of 24.4% during 2020-2027.

But what really is smart clothing?

Smart clothing, like all other forms of mobility, is when technology meets textile. With the use of conductive textiles and advanced analytic systems, smart clothing allows the people who wear it to capture and quantify data from their bodies and mind in real-time. Just like every other technology that has the term 'smart' associated with it can sense and respond in a controlled or predicted manner to environmental stimuli. However, the degree of "smartness" varies from device to device.

The clothing industry has reaped the benefits of these technologies with the rapid commercialization of this industry.

If you are into technology and all the advancements that come with it, then you must be familiar with wearables, such as fitness bands, also known as an activity tracker, that track your steps and calories burned. Today, the limits of textiles are being stretched even further, and understandably smart clothing is the next step in wearable health and wellness consumer products.

On the outside, when it comes to appearances, smart clothing is designed to look like any other garment you have in your closet. However, the technology lies within the fabric. Fibers, yarns, fabrics, and other wearable materials have been successfully developed for technical and high-performance delivery to the user who wears them. Unique from other wearables on the market, smart clothing doesn't require additional accessories to charge and operate. Smart clothing is hassle-free, with no extra effort required.

Wearable sensors embedded in clothing are improving to not only become the most wearable but also to become the least disruptive to the wearer in terms of convenience and comfort yet. They want you to gain the benefits of this wearable technology just like you would in any other device, without having to change any of your behaviors.

5.4.1 Uses of Smart Clothing

Smart clothing garments will provide users with insightful real-time readings to effectively keep a check and manage their health and wellness. With their inherently responsive materials, they can.

- Automatically track the training progress and monitor the wearer's physical state

- Continuously enhance your workout efficiency and improve performance based on the results.

- Prevent training, related injuries, and work out overload by notifying you.

And, with these effective interactions, health check-ups, and monitoring results that it provides, many believe that it can renovate traditional medical facility-based services.

5.5 Sports Gadget ISNs

Fitness technology has come so far. With the rising popularity of workouts and being healthy, sports gadgets are another popular item. Unlike smart clothing, sports gadgets are not attached to a garment. The advancements in sensor technology are now such that the size of the device used to 'sense' is significantly reduced to fit in your wrist, giving you a comfortable application. They are a device present to provide a solution with its use of flexible sensors. And, just like a mobile phone, generally connected to other devices that you can either carry with you or move around in depending on your needs.

With exceptional health benefits, advanced fitness technology allows you to analyze most, if not every aspect of your sports or workout routine. It allows you to monitor your progress, stabilize your strengths and improve upon your weaknesses. Currently, there are a plethora of smart fitness gadgets available in the market today for your sporting needs.

5.6 Parking ISNs

Parking a car is far from ideal. It is without a doubt a painstaking process where each car owner wants the best spot, and also, they want to keep their vehicle safe. This is the reason why parking sensors were invented. They have helped ease and, in some cases, eliminate this tiring process of parking. These parking sensing devices, also known as proximity sensors, are located on the bumpers of a vehicle in order to assist the driver when parking.

How do they work[9]?

As the name suggests, the proximity sensors work by measuring the proximity of your car to an approaching object. Depending on where your parking sensor is attached, it can either sense the object that is in front or behind and alerts the driver when they get too close to minimize risk. These proximity sensors are equipped with a special alert-like tone and, every time the driver comes close, it produces that beeping tone. For a better understanding of your surroundings, the alert gets faster and more frequent as the car comes to the object.

There are currently two types of parking sensors on the market, ultrasonic and electromagnetic, and while they both do the same thing, they do so in different ways.

Ultrasonic Sensors

Ultrasonic sensors work while the car is motionless or while it's moving, and it uses sound waves to detect any object that is in front of the sensor by pulsating sound at a high frequency. This type of sensor reflects off nearby objects so a receiver can catch the reflected waves and calculate the distance to the detected objects. Ultrasonic sensors either emit a sound to alert the driver or translate the warning to simple writing, a visual, or any graphic that uses color to represent the vehicle and objects.

However, there are disadvantages to this system as an ultrasonic sensor may not be able to detect objects that are:

- Too thin to reflect sound, e.g., posts and bollards

---

[9] Viktor Gubochkin, How To Use IoT For Smart Parking Solution Development, April 14, 2020.

- Too low to be detected, e.g., the footpath

- Too flat, as they won't reflect sound back as well

- To the side, so any poles that are on either side as ultrasonic sensor only pick up objects directly in-front

Furthermore, ultrasonic sensors are affected by interference which means if anything is obscuring the sensor, it will, unfortunately, affect how the sensor reacts.

Electromagnetic sensors

Unlike ultrasonic sensors that are restricted in their use and can only detect objects in front, electromagnetic sensors can detect objects nearby the entire vehicle. The Electromagnetic sensors transform a quantity to be measured; in this case, it uses frequencies to create an electromagnetic field that can detect anything that enters it. As a result, this means that electromagnetic sensors don't work when stationary unless the objects themselves are moving, e.g., pedestrians or other cars.

While these sensors may not work as well while stationary as they do in ultrasonic sensors, electromagnetic sensors don't suffer the same interference problems as ultrasonic sensors. They are far more reliable when it comes to their surroundings, and they are also more sensitive, which allows them to pick up objects that ultrasonic sensors can't.

However, while they don't have any apparent disadvantages, they are usually more expensive than their ultrasonic counterparts and may not be suitable if you are someone on a budget.

Challenges

Even though parking sensors are great assistance tools that significantly ease the task of parking, they can be considered lacking to some. As such, some manufacturers offer alternative assistance like a reverse camera that works in isolation or in conjunction with parking sensors to greater simplify and improve the overall parking experience.

5.7 Putting it all together

In this portion, we have emphasized the importance of high-efficiency and reliable transmissions when it comes to building a mobile, more secure future. Mobility is one of the most important aspects of our future and whether it's to improve the quality of life for citizens by giving them sensor-loaded mobile phones or whether it's providing them with the technology to monitor their own health with a portable gadget.

However, it is not to say that the industry is without its limitations. We talked about how these mobile devices are causing users to be a bit more skeptical about their privacy, and with that, trust issues are emerging. On the one hand, their information like location, identity, mobility patterns, and social connections is easily accessible by these devices in order to provide a better user experience. On the other, the constant scare of this date being susceptible to any leaks is slowing the growth of this sector and is emerging as a challenge. Therefore, for their success, it is necessary to come up with trust-based strategies and provide the user with enough knowledge so they can be sure the devices they use are reliable and secure in their communications.

- Research efforts in the field of sensors and mobility had improved urban traffic control systems.

- With the integration of information technology, cloud, and social network services, these devices have become more efficient to enhance their quality of life.

- A smart clothing system possesses machine intelligence to read, monitor, and analyze our real-time data through a perfect combination of sensors, mobile clouds, and smart clothing.

- Parking sensors are an innovative concept. They rely on sensors and minimal manual user input so they can allow the driver to effortlessly park their cars.

Chapter 6: ISNs and Our Energy Sources

"If you want to find the secrets of the universe, think in terms of energy, frequency, and vibration."

Nikola Tesla

"Energy is liberated matter, matter is energy waiting to happen."

Bill Bryson, A Short History of Nearly Everything

Nowadays, the overwhelming rate of urbanization, as well as overpopulation, has brought many global concerns, one of them being in the realm of sustainable energy production and access. Understandably energy is required to make a lot of things work, and the source for anything depends on how we acquire it. In the case of energy, we acquire it from an inventory of sources. Our energy supply comes mainly from fossil fuels, with some nuclear power and renewable sources passing out the mix of all. However, with our standard of living progressing at an unprecedented scale in human history, we can understand that an energy transition has been underway for the past two decades. To a larger extent, this change and its success in recent times can be attributed to our increasingly sophisticated uses of energy due to the electrification of all things.

If we put our minds to it, our energy uses depend heavily on what era we're in. Today the strength of this industry ranges from the speed of transportation, education, communication, the innumerable comforts and conveniences of home and workplace we use energy in, and the security of the nation all derive from ever more practical limits to energy, resourceful provision, and application of various sources and forms of energy.

6.1 Background and Introduction

Today, the energy sector is highly reliant on fossil fuels with large carbon footprints, instituting nearly 80 % of final energy worldwide. During the initial years of the development of IoT, energy was not given the importance it deserves. The IoT devices were typically powered by non-renewable energy sources and were assumed to have limitless potential. Different batteries were thought to be adequate for powering the devices that were used. However, the idea has proven to be not only unsustainable but also unfeasible for the long term as batteries are thought to have a limited lifetime and are often impractical and expensive to replace. And not to mention the heavy environmental damage disposed of batteries can cause.

Therefore the trend had to change to provide energy efficiency, i.e., consuming less energy but delivering better service with the help of technology is the correct alternatives to diminish our adverse impacts of unsustainable energy use. The evolution of IoT and electrification of all things, including new forms of cryptocurrency, has required stable and reliable power sources. With IoT, the use of smart devices in places where these batteries cannot be charged became prevalent. And, as the IoT encompasses a wide range of technological domains, it is fair to call it the best possible outcome for energy and its sources. In this section, we will talk about ISNs and the role it plays in the multiple energy sources of today and how IoT helps us in efficient power generation, transmission /distribution, real-time demand/peak monitoring, energy regulation, and all things energy and technology.

6.2. ISNs in electric power generation/distribution

The age of equipment in the power sector and poor maintenance problems cause a lot of unreliability and safety issues. It can lead to a high level of energy losses and irregularity that only disrupts the energy process. We have seen more blackouts in the last decades than before. Equipment is sometimes more than 40 years old, hefty on the pocket, and not easily replaceable. This is when IoT comes into play; it can eliminate or lessen most of these challenges in the management of power plants. By applying IoT sensors, Internet-connected devices can forecast and real-time flag any failure in operation or abnormal decrease in energy efficiency, alarming the need for maintenance. This increases the reliability and efficiency of the system, in addition to reducing the cost of maintenance.

IoT sensors have started catering to generation, transmission, and distribution equipment when it comes to energy. IoT employs sensors and communication technologies that enable energy companies to monitor all three aspects from generation, distribution, to transmission, all remotely. These sensors measure parameters such as vibration, temperature, and wear to optimize maintenance schedules. This preventative maintenance approach can significantly improve reliability by keeping equipment in optimal performance and providing the opportunity to make repairs before it fails. In addition to supporting preventative maintenance programs, this technology enables virtual troubleshooting and support from monitoring processes during manufacturing. However, in terms of monitoring developments during manufacturing, IoT and its enabling technology play an essential role in shaping this industry and generating energy.

Even in the early stages, IoT contributed to the power sector in remote locations by monitoring and controlling equipment and processes, alleviating the risk of loss of production or blackout, and using efficient energy remotely. Today, IoT is employed in different sectors to design a fully connected and flexible system to reduce energy consumption while optimizing production. The IoT-based digitalization source transforms the energy system from a unidirectional cycle, i.e., from generation through multiple energy grids to consumers, and later sends it off to an integrated energy system. Different parts of an integrated smart energy system are depicted in Figure 6.1.

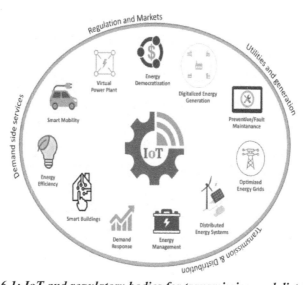

Figure 6.1: IoT and regulatory bodies for transmission and distribution

As the figure 6.1 shows, there are different bodies of regulations and corporations that handle each stage of the energy cycle. IoT aids energy systems to enable sharing of the data by all parties involved for better communication and utilization of resources making improvement in regulations a reality. For example, nowadays, many homes and buildings are powered by solar or wind. These energy sources are not allowed to be off the grid and push the energy to the grid for the utility companies to manage the peak loads better. Hence storing energy for night use or when the wind is not available poses a challenge to the utilities.

However, battery-backed solar/wind energies are now allowed and enabled by IoT devices to not only store energy but also provide the necessary power while being on the grid and allowing correct metering by the utility companies. These batteries nowadays consist of lithium-ion batteries, and in the future will transition to either liquid metal batteries or redox batteries as these new technologies tend to be safer and do not have the over-heating problems of lithium-ion batteries while being IoT enabled and provide a much more efficient means of storage and local transmission of energy.

In energy systems, the major use and struggle of IoT platforms are to monitor and help save energy use while multiplying the benefits. This essentially means that IoT systems are collectively trying to increase our energy efficiency by saving energy and distributing it more efficiently. In energy systems, in order to allow communication using IoT, a massive number of IoT devices need to be set up to transmit data. To run the IoT systems and transmit that massive amount of generated data, a considerable amount of energy is needed[10]. Therefore, the heavy energy intake of IoT systems remains a critical challenge that needs to be tackled. However, with time as many people have started to gather this as a problem that needs solving, numerous approaches have tried to reduce the power consumption of IoT systems. For example, by setting the sensors to have an idle or sleep mode, we can direct them to go to sleep and just work when we need the device to work. This feature not only saves power consumption but also reduces the amount of data that needs to be gathered and transmitted to do the job, further reducing the energy required to do the job. For this to work, manufacturers have designed efficient communication protocols that will allow distributed computing techniques and enables energy-efficient communications.

An example of such a manufacturing standard is Industry 4.0. Industry 4.0 integrates advanced production and operation techniques with smart digital technologies, enabling an interconnected network that will allow for greater knowledge sharing, as well as improve operational efficiencies. It represents the movement to integrated, advanced production and operation techniques with smart, digital technologies to create an interconnected network that will allow for faster data sharing and analysis. The result of Industry 4.0 yields a decentralized decision process that allows for real-time data and analysis to drive decision-making, moving away from traditional linear data sharing.

A process known as the physical-to-digital-to-physical (PDP) loop, which allows for the interconnectedness between physical and digital technologies.

---

[10] Kaur, N.; Sood, S.K. An energy-efficient architecture for the Internet of Things (IoT). IEEE Syst. J. 2015, 11, 796–805.

An integrated system would reduce operational costs, increase efficiency, allow for a continuous rise in technological innovation, improve product agility, and increase plant agility. While Industry 4.0 is most associated with manufacturing, smart technologies can transform how parts and products are designed, made, used, and maintained, and its PDP loop amplifies the data collected throughout power plants, enabling a continuous and cyclical flow of information between physical and digital worlds. Processing mass quantities of data and using, processing, and sharing it simultaneously is next-generation technology that sensors and controls can help enable.

The future fossil energy domain is transforming to a more secure, integrated controls/optimization, storage, and modern infrastructure, as shown in Figure 6.2 below.

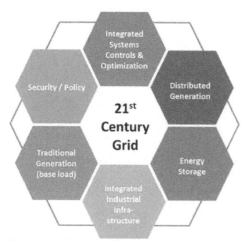

Figure 6.2:. Showing the components of the 21st Century power Grid

Figure 6.2 shows not only energy storage element being part of the domain but also distributed generators along the side of traditional generation overarched by security and policy. Security on intelligent power domains is must-have, which includes not only the physical security but also the cybersecurity of the domain. Cybersecurity also plays an important role in the development of successful sensors and controls. As technology can bring more and more pieces of the power plant online, it is important to develop sensors and controls that maintain and upgrade the security of the plant in order for the infrastructure to continue to provide a secure and reliable source of energy. The shift toward a smart grid will add thousands of smart devices to power plant operations that will need to prepare for enhancing the cybersecurity capabilities of both IT and OT systems. Not only will devices be added to increase the plant's ability to operate online, but changing cyber technology, including blockchain, can be used to further change the energy landscape and utility sector.

Power generation or electricity generation is the process of generating electric power from sources of primary energy such as heat (thermal), wind, solar, and chemical energy. For decades, sensors have provided the power generation industry with innovative sensors and control solutions for challenging applications. These applications include but are not limited to

- Fuel Cells

- Generators

- Power Systems

- Turbines

- An example of ISNs in power plant usage...

https://www.pcb.com/applications/energy/information/sensors-to-monitor-power-plant-machiner

Further examples of an ISN for power distribution can be found in

https://www.osti.gov/servlets/purl/1004093

6.3 ISNs in water dams

Dams store water, provide renewable energy, and prevent floods. Unfortunately, they also worsen the impact of climate change. They release greenhouse gases, destroy carbon sinks in wetlands and oceans, deprive ecosystems of nutrients, destroy habitats, increase sea levels, waste water, and displace poor communities. In order to optimize and reduce the negative impact of the dams, better monitoring through ISNs can provide the necessary information in order to control some of the effects. Three very important parameters that can be monitored sensed, and decisions made to reduce environmental impacts are water level, water quality, emissions from the dams.

Water level monitoring can be done using inclinometers, radars, ultrasonic sensors, pressure sensors that are locally networked or cloud based.

Water quality monitoring is defined as the sampling and analysis of water constituents and conditions. These include pollutants, such as pesticides, metals, oils, and constituents found naturally in water that can nevertheless be affected by human sources, such as dissolved oxygen, bacteria, and nutrients. The magnitude of their effects can be influenced by properties such as pH and temperature. For example, temperature influences the quantity of dissolved oxygen that water is able to contain, and pH affects the toxicity of ammonia. Water quality monitoring will typically use pH, KH or carbonate hardness, temperature, chromophoric dissolved organic matter (CDOM), fluorescent dissolved organic matter (FDOM), turbidity, phosphorus, nitrates, dissolved oxygen, conductivity, and total dissolved solids (TDA) sensors. Many companies provide pieces of this puzzle, but an integrated monitoring system is needed to sense, analyze, and report the conditions and quality of water dams in almost real-time in a secure fashion. Currently, the process is manual, labor-intensive, and always time-consuming. A good example of the next-generation integrated system using ISN is given in Smart water quality monitoring system with cost-effective using IoT" by Sathish Pasika*, Sai Teja Gandla[11].

---

[11]Pasika, S., & Gandla, S. T. Smart water quality monitoring system with cost-effective using IoT. Heliyon, 6(7), (2020).

As can be seen in Figure 6.3, the block diagram includes the pH, Turbidity, and DHT-11, ultrasonic sensors powered locally, collecting data with a local MCU, Processed and pushed to do the cloud.

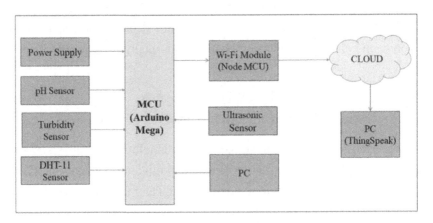

Figure 6.3:The system software flow diagram.
Although this advanced system, there still are other factors that are not included for sensing and can be done so.

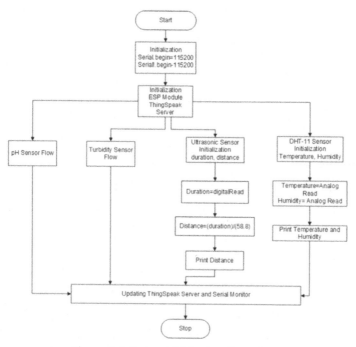

Figure 6.4: Sensorial Monitor flow chart

6.4 ISNs in mining

ISNs use is already widespread in mining, including in applications such as automation and remote operation and data analytics for control and optimization. Some of the present sensing technologies outline upcoming sensing services in mining.

New and emerging sensing technologies against the current challenges faced by the mining industry, including localization and tracking, imaging, 3D ranging and mapping, machine and equipment condition monitoring, composition measurement, exploration, health and safety, and tire-pressure monitoring.

A variety of sensor technologies today concern the detection, tracking, and communication of relative and absolute positions of humans but also vehicles, equipment, and other resources. In mining, such technologies are of significant interest, mainly in driverless vehicle control, remote operation of equipment and asset management and tracking, site security, and personnel location. The main types of such sensor systems are satellite (GPS) and terrestrial (radio frequency-based) positioning systems, dead reckoning systems (inertial sensor-based position tracking), beacon sensors, and tagging technology.

Autonomous vehicles in Mining

The mining industry has been employing driverless vehicles for several years, enabling cost reduction and safety improvements in transport. Local sensing and control have provided increased autonomy to vehicles, which can now operate continuously and feature various supporting features such as route and location optimization, collision avoidance, and maintenance control. Recently, the combination of local (proximity) and remote (positioning) sensing has led to increased reliability of such systems. Capabilities such as real-time machine tracking, scheduling, assignment, and productivity management are also becoming available. Anticipated developments in this area mainly derive from the rapid development of Advanced Driver Assistance Systems (ADAS) for passenger cars.

These systems provide an increasing range of functions, including navigation, collision avoidance, parking assistance, and self-parking, lane change assistance, and adaptive cruise control. Because of the variety of requirements, these systems also integrate ultrasound, Radar, Image Sensors, Infrared sensors, Ultrasonic sensors, and LIDAR sensors. Mining automation is expected to benefit from the enhancement and cost reduction of these systems in the automotive market, including not only sensors but the related software for functional integration. The result will be a much richer mix of sensor data from vehicle automation and more precise and reliable control. The availability of fleet-level data will also allow high-level process optimization, taking advantage of data analytics for vehicle maintenance and supply chain management.

Real-time global positioning

This has been supporting localization and tracking in mining for decades. A terrestrial receiver can determine its position from a known code sequence coming from four or more satellites of known location, using the synchronization-determined time of flight and the time of transmission and satellite location, which are also included in each signal. The two systems currently in service are GPS and GLONASS, each providing an accuracy of around 5 m.

Currently, they are becoming more accurate, mainly by employing secondary receivers in fixed locations. A commercial example of such hybrid systems is the assisted GPS service found in most smartphone products. Also, the Chinese BeiDou and European Galileo systems are expected to provide cm-scale absolute positioning as a commercial service. It is expected that cm-accurate outdoor global positioning will be used routinely in mining operations within 3-5 years. For indoor locations, local implementations of RF positioning are available based on a similar working principle.

In recent advances of such systems, an ultra-wideband real-time tracking system has been adopted by NASA and proposed for various applications, including navigation in mining sites, with features such as high precision, scalability, and (optional) dual communication. These systems are often able to estimate position from the last known position and kinematic state and subsequent estimation of displacement by local tracking of acceleration (accelerometer sensor) and orientation (magnetometer and or gyroscope sensors). This technique is known as dead reckoning. Combined RF and inertial sensor systems for mining applications have already been utilized in cars and airplanes.

Purely inertial positioning systems are also possible, especially for applications where a known checkpoint can be visited regularly. As underground operations become more automated, indoor positioning with off-line capability will become increasingly important. Therefore, it is anticipated that the use of dead reckoning in mining will increase greatly in the following few years.

Further possible applications of such positioning systems could include automatically operating shovels and robotic arms. An example of such an initiative can be found in the recent Scanning Range Sensor program of CRC mining, where range sensors are evaluated on-site for mining shovel control. Beacon and RFID sensors Location beacon systems are currently used in mining monitoring for cave tracking. Ore locations are marked by beacons to monitor the flow during mining activities in order to predict waste ingress into the ore. This real-time cave motion monitoring technology can, in turn, enable cave design optimization. The current state-of-the-art involves large-scale beacons emitting a rotating magnetic field. Their location can be determined by a stationary detector. Beacon systems can also be used for local positioning by providing proximity information, checkpoints, or fencing of particular areas. An example of such a system can be found in the introduction of RF beacons into consumer electronics wireless devices for indoor positioning.

In these applications, low power operation and energy efficiency are key requirements, and hence recently developed communication protocols, such as Bluetooth Low Energy, have been adopted. A similar positioning principle can also be achieved using RFID tags. A moving object can be tracked by a local RFID reader, at a range of several meters for passive tags or at a range of several tens of meters for active (battery-powered) tags. Inversely, a moving object equipped with an RFID reader can track its own position by identifying passive or active tag checkpoints of known locations. RFID-based beacon solutions have the primary advantage of allowing very large numbers of tags at a minimal cost.

Asset tracking and position information

Proximity sensors can also be used for positioning and tracking, either as stand-alone systems or in tandem with other positioning technologies. They are usually short-range sensors, and their current use in a mining operation is mainly related to the position and orientation control of automated equipment such as driverless vehicles, including collision avoidance and personnel security. In orientation control, proximity switches and scanning systems operate alongside central positioning systems (terrestrial RF, GPS, etc.), offering complementary information that enhances accuracy, speed, and reliability in related automation. In collision avoidance, proximity sensors are used to detect and prevent dangerous situations such as unsafe machine-to-machine and machine-to-personnel distances and unauthorized presence.

The main proximity sensor technologies are based on inductive, capacitive, electromagnetic, ultrasonic, and optical operating principles. The inductive and capacitive proximity sensors detect the disturbance of a magnetic and electrostatic field, respectively, by the close presence of an object. Therefore, they operate in proximity, typically below 100 mm. Electromagnetic, ultrasonic, and optical proximity sensors offer long-range detection and could potentially provide further critical. An example of ISNs used in IoT-enabled mining is shown in figure 6.5.

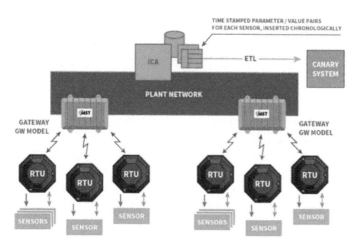

Figure 6.5: Plant network

Health monitoring in Mining

With the rapid growth in the healthcare sector, sensor data offers significant changes that will impact the entire process of how the healthcare industry operates. Medical institutions and healthcare providers see an opportunity in this fast-paced transformation and are using sensors to collect large amounts of data that will prove to be useful. The real-time data they collect on their patients help in providing insights on a patient suffering from a variety of diseases. And, for better monitoring and understanding, health care professionals are making a choice to organize this data to form a record or database of sorts.

However, this rapid data collection also introduces a new problem known as the data overload problem. This means that the data the system has collected can either be faulty or completely irrelevant. To avoid that from happening, both systems and other interested parties need to complement the data that they acquire from their sensing capabilities with data mining so they can easily transform and consume the collected data for meaningful intelligence as shown in figure 6.6.

Health monitoring in mining can essentially be divided into two sections[12].

1- Clinical Mining for both in and outpatients.

2- Non-Clinical Mining to analyses wellness management, activity monitoring, the use of smart environments (e.g., smart home scenarios), and other machine-sensed environmental data concerning human social behavior known as reality mining.

---

[12]Sow, D. et al. "Mining of Sensor Data in Healthcare: A Survey." Healthcare Data Analytics (2015).

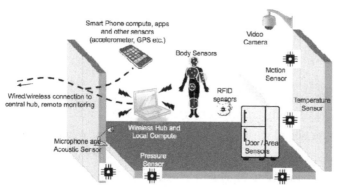

Figure 6.6: collected data for meaningful intelligence

6.5 ISNs in Harvesting Energy

Energy harvesting devices convert ambient energy into electrical energy with applications in industrial, military and commercial markets. Some systems convert motion, such as that of ocean waves, into electricity to be used by oceanographic monitoring sensors for autonomous operation. Future applications may include high power output devices (or arrays of such devices) deployed at remote locations to serve as reliable power stations for large systems. Another application is in wearable electronics, where energy harvesting devices can power or recharge cellphones, mobile computers, radio communication equipment, etc. In mining, the RF tags are energized when passing through an RF field and do not have any other power sources.

Energy can also be harvested to power small autonomous sensors such as those developed as a chip-based technology known as MEMS sensors. These systems are often very small and require little power, but their applications are limited by the reliance on battery power. Scavenging energy from ambient vibrations, wind, heat, or light could enable smart sensors to be functional indefinitely. Typical power densities available from energy harvesting devices are highly dependent upon the specific application (affecting the generator's size) and the design itself of the harvesting generator. In general, for motion-powered devices, typical values are a few μW/cm for human body-powered applications and hundreds of $\mu W/cm^{13}$ for generators powered from machinery. Most energy scavenging devices for wearable electronics generate very little power.

6.6 Putting it all together

Our energy supply comes from a lot of sources, and they vary in their kinds depending on where we acquire these energies from. Whether it's solar energy or renewable energy, the world depends on one or the other to carry out its activities. But in order to generate, store, improve, distribute and use this energy for the better, we need technologies that employ sensor and communication technologies.

This is where the Internet of Things is helping systems drive this transformation. Looking back at the chapter, here's a quick summary of how IoT is transforming the energy industry:

- IoT sensors are helping improve and increase the efficiency of the production and distribution of electricity. Real-time data extracted from these sensors help balance production, management with the life cycle cost of maintenance and running of equipment that later on enables fast computations.

- Monitoring and managing dams is important for the regulated flow of water and energy. However, human error can cause irreversible damages to the dam's structure and the flow and production of energy; hence, monitoring the dam's safety and water management is extremely important considering both the situations like water scarcity and excess of water.

- ISNS technology has re-engineered processes and activities within the mining industry for better mobility: Cloud computing, Sensing,

---

[13] WAREHOUSES AND COLD STORESCOMPREHENSIVE MONITORING OF TEMPERATURE AND HUMIDITY. (2021). Monitoring Temperature and Humidity. https://www.efento.gr/en/application/food-industry/monitoring-temperature-and-humidity-in-food-warehouses-and-cold-stores/

Analytics, and Security. The real value of implementing digital technologies for additional useful, productive, and connected mine can be seen in many industries around the globe. From mining that data of automated vehicles to the health care sector, the mining industry has been using sensors to be more accurate and productive.

- To harvest energy from the sustainable power supply, many sources such are mechanical motion, radiation, thermal gradient, and light can be used. For every ideal energy harvester, the main goal is to provide constant energy. However, the many limitations can be optimized by the management and energy storage provided by the innovative design and solutions of IoT devices. It is a promising field that allows and supports IoT devices the ease to generate electrical energy by absorbing energy from the environment.

Chapter 7: ISNs and Our Freedom

"National security always matters, obviously. But the reality is that if you have an open door in your software for the good guys, the bad guys get in there, too.

Tim Cook

"When you give everyone a voice and give people power, the system usually ends up in a really good place. So, what we view our role as, is giving people that power."

Mark Zuckerberg

7.1 Background and Introduction

ISNs and IoTs in the digital age enable us to collect data from every aspect of our lives. From home to our mobile devices, to factories, buildings, environments, industrial complexes, automobiles, and roads. These devices make us more knowledgeable about our needs, surroundings and can help us make data-driven decisions at our will. Although this gives us more peace of mind, it cannot come at a cost to our privacy.

Our personal data should be ours, and we should have the right to choose who accesses this data. Also, as states and corporates continue to maximize data collection and retention, governments worldwide must take concrete measures to address existing and emerging threats to affirm and safeguard the fundamental human right to privacy – our reputation and freedom. There have been series of resolutions that progressively elaborated international standards on how to secure the right to privacy online, personal data, including contending with the risks posed by new and emerging technologies. On December 16, 2020, the U.N. General Assembly adopted a new resolution on privacy in the digital age. The resolution, which was co-led by Brazil and Germany and co-sponsored by a record cross-regional group of 69 countries, reaffirms the fundamental importance of the right to privacy and renews international commitment to ending all abuses and violations of this vital right worldwide. Although a concept at best for now, we are all striving to reach the stage that these privacy is protected.

This resolution builds upon previous resolutions of the same title, with new language on equality and non-discrimination, artificial intelligence, biometrics, encryption, and gender, as well as addressing privacy challenges in the context of the COVID-19 pandemic. However, the resolution missed the opportunity to provide firm and strong recommendations to states on some emerging threats for the right to privacy, such as on facial and biometric recognition technology. It is crucial that future versions of the text continue to respond to the key issues of the day firmly.

There have always been concerns on miss-use of digital data for:

Equality and non-discrimination Indivisibility of human rights. Use of Artificial Intelligence and machine learning in data mining Biometric technology.

The Internet of Things is a new paradigm that is revolutionizing the world of computing; billions of objects will be connected to the Internet, which is a very powerful tool for marketers to understand the market better and increase the effectiveness of advertising campaigns. With the advent of crowdsourcing, companies are increasingly opting for a collaborative approach to better present their products and monitor people's consumer habits (e.g., Amazon's Alexa), but with the large amount of data sent by its connected terminals, the problem of collecting, storing, analyzing, and retrieving these data to better orient their marketing strategy arises. In the context of these trends, we need to be aware of technologies that also can safeguard our information and data. These include encryption algorithms, including but not limited to blockchain and quantum encryption/decryption.

7.2 ISNs helping us to free our time and resources spent on the mundane

Companies historically invested in technology for the sake of the technology. With the assumption that computerized systems were always more efficient. However, companies now demand to know the savings from all capital investments, including computer systems. Today, virtually all automation must be financially justified. For example, a recent pull-out section in the Wall Street Journal on technology opens with: "It's the morning after. For the past ten years, companies have been on a blind-faith buying binge investing well over $1 trillion in new computer systems to embrace the future and gain a competitive edge. Now, many of them are awakening with a hangover and wondering: What was it all for, and where did we go wrong?" *

"Automated data collection technology is used to enter information into a business computer system. It relies on machine-readable bar code symbols to increase the speed and accuracy of collected data." *

* *The Wall Street Journal, Technology – Unleashing the Power, Dennis Knearle.*

Calculating the current cost of the tasks that you plan to automate is critical.

The good news is that sensor systems really do save lots of money. Even skeptical, tight-fisted operations managers can understand the benefits of eliminating manual data entry and errors. The benefits of being wireless is intuitive to an experienced warehouse manager.

Change – even positive change – causes uncertainty, so make allowances and be conservative on productivity gains from automated data collection. This generally fit into two categories: direct savings and indirect benefits. Direct expenses typically include labor, material, operating expenses, inventory, and fixed Asset reductions such as reduced equipment needs or reduced equipment losses due to better equipment tracking/maintenance. The indirect benefits examples are reduced out-of-stock situations and up-to-date product availability information. General Efficiency improved employee morale by automating boring tasks such as data entry and creating more enriching job opportunities for workers. More Satisfied Customers due to higher quality, faster response, and improved accuracy.

The most common area of savings from automated data- collection is labor costs. When calculating labor savings, determine the variable cost of labor. The variable cost is the cash expense that varies in direct proportion to the hours worked. Variable costs generally include wages, payroll taxes, and employee benefits. A second major category of savings is inventory. Warehouse managers will admit they would reduce inventory if they had better, quicker, and more reliable information about what's in stock at any given time. Thousands of companies have used barcoding to improve inventory systems and give management better information.

We all know that reducing inventory saves money. As with labor, inventory savings should focus on the variable holding costs of inventory. When quantifying how much your inventory application will save, determine the variable holding costs of your inventory. Many companies have a rule-of-thumb they use to calculate the cost of carrying inventory. Depending on the industry, these costs range from 15% to 35%.

Intangible benefits are important to your company and, therefore, should be highlighted in your proposal. At the same time, you may not want to quantify them. It all depends on the culture of your organization and the nature of the key decision-maker. If the decision-maker is a "by the numbers" type of person, you probably should be painfully conservative with your assumptions and estimates of soft benefits. There are ways to quantify soft costs if you believe it is necessary. The risk is that the reader may disagree with your assumptions or method and, instead of focusing on the benefit of the bar code system, criticize your approach. If you decide to quantify some intangible benefits, try to use conservative assumptions that are validated by well-regarded people in your organization.

A good example can be found in a research article called **Comparison of Manual Versus Automated Data Collection Method for Hematological Parameters**[14]

On one side, from giving us a life of comfort and ease, these devices also provide cost-saving and time-saving benefits. These countless benefits are among the main drivers for intelligent technology adoption, particularly for governments, healthcare, infrastructure, and businesses. The information collected by IoT devices/smart objects can automatically control the situation in self-managed autonomous systems. Once the data collected is shared with the other entities in proximity, they can make some decisions of their own, cut back our precious time, or send through communication hardware to the intelligent design in the cloud.

Apart from saving our time with driving, health care, and food control, the use of IoT has a vital influence on marketers and advertisers since it provides them with access to big data they can rely on. Marketers can track and record products, their sales, calculate the number of customers daily, analyze purchasing behaviors and understand the individual applications of products. The calculated outputs will save time and eliminate the need for surveys to collect only costly and time-consuming data with IoT, where ideas can be collected from the actual use of connected products and related data. It can also improve direct and hyper-local marketing, enhance the quality of their services, and add new features by observing customer responses using IoT devices.

This easy access to data can directly help all forms of marketing since the information is first-hand. The conservative marketing techniques that were once used are very costly, and it's no secret that it took more time. But with the emergence of smart technologies, the dissemination of this data can reach more individuals regardless of their demographic, psychographic, or

---

[14] Ayyakkannu Purushothaman, Comparison of Manual Versus Automated Data Collection Method for Haematological Parameters, February 19, 2019.

geographic generalizations. The increase of IoT will influence all marketing and advertising companies, particularly those focused on big data analytics and ones that use data to make future decisions. With more extensive data from consumers and businesses which became quickly available to marketers, analysts can turn raw data into valuable insights, recommendations, and predicted outcomes.

Due to the significant role of IoT in enhancing services in today's world, it's seen as a key that unfastens hasten activities. According to Statista,[15] the total installed base of Internet of Things (IoT) connected devices worldwide is projected to amount to 30.9 billion units by 2025, a huge jump from the 13.8 billion units at the time of this writing in 2021. Whereas the entire IoT market reached only $100 billion in revenue in 2007, forecasts suggest that this figure will grow to around $1.6 trillion by 2025.

With that said, it is important to note that IoT is an extensive scale network consisting of diverse constrained devices and smart objects. Such a constrained network imposes a significant impact on the designing of different protocols[16] . However, taking this into account generally enables for designing a broad set of standards and protocols. These protocols are supposed to offer efficient and scalable communications and allow developing and deploying applications/services adopted for various environments that late on cut our total costs.

[15] Lionel Sujay Vailshery, IoT and non-IoT connections worldwide 2010-2025, Mar 8, 2021

[16] A S Abdul-Qawy et al. Int. Journal of Engineering Research and Applications www.ijera.com ISSN: 2248-9622, Vol. 5, Issue 12, (Part - 2), pp.71-82, December 2015.

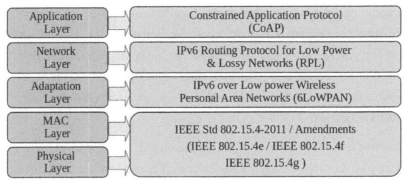

Fig 7.1 Emerging Protocols for the IoT

7.3 ISNs reducing time to find locations, objects

Whether it's for personal use, for organizations, or for the government, location is the key that brings context. IoT has a subcategory known as the "location of things." Location is a necessary dimension of the IoT concept that comprises "things" to sense and communicate their geographic position. In this context, the location simply acts as a search engine for geographic data. Location acts as an organizing principle for anything that is connected to the Internet.

The start of location of things started with GPS[17]. In the 90's we were introduced to the GPS. However, there are lots of our devices and sensors — along with other objects where GPS has no actual reach. That's where indoor positioning systems (IPS) became the next big thing inside the "location of things." With IPS, the location data gathered can help everything from discovering devices and supplies, equipment, step-by-step navigation of indoor spaces such as someone's residence, shopping malls, assisting with logistics in warehouses, locating stations, enabling geofencing around sensitive data, assisting in social interactions, and more. This is how applications like Google maps predict our location. They use location-based technologies with the help of a single line of code, any device or sensor

---

[17] Olutosin Taiwo, Absalom E. Ezugwu, Smart healthcare support for remote patient monitoring during covid-19 quarantine,Informatics in Medicine Unlocked,Volume 20,2020.

dispersed across a network (*an IoT edge device*) can access Google's geospatial database of Wi-Fi and cellular networks using the Google Maps Geolocation API.

We must understand that as more and more "things" are connecting to the Internet, the amount of data on our location coming in through the use of IPS is overwhelming. Day after day, an uncontrollable wave of new data is becoming part of the location of things. Data from various sources, some online and some held up or being shared in multiple clouds is always being circulated, and it will only become more complex. IDC predicts that by 2022[18], over 90% of enterprises worldwide will rely on a mix of on-premises, legacy platforms, and public and private clouds to meet their infrastructure needs. We need filters to gather the information that is relevant for us. With the advent of these devises companies, individuals no longer must waste precious moments to improve their inventory management, track product usage, and monitor selling rates and locations. But this easy access to devices that lets them know of our where about raises a lot of security concerns and risks to overcome that IoT devices have tried to solve.

7.4. ISNs helping us feel safe

The Internet of Things (IoT), without a doubt, has added to our overall safety. I call it predictable safety through sensing. We live in a world where we can efficiently use technology to prevent bad things from happening, and that is exactly what the complex technology of ISNs and IoT Devices have hoped to achieve.

Internet of Things (IoT)-enabled emergency systems that help eliminate human error and resolve issues have made many areas safer. For example, cities, schools, health care systems, airports, and security at home pose quite a few security challenges that IoT has helped solve.

---

[18] Bruno Aziza ,"What Customers Teach Us About The Future Of Data," Forbes. Jun 3, 2021. https://www.forbes.com/sites/googlecloud/2021/06/03/what-customers-teach-us-about-the-future-of-data/

The basic technology that IoT uses to detect nearby dangers includes:

- Sound sensors that can detect loud noises, such as explosions or gunshots,

- IoT-enabled light bulbs that can alert when an intruder breaks in,

- Lights to indicate danger or safe zones,

- Irregular data can trigger real-time incident reporting,

- Biometric identity management solutions are used in airports and other areas where security is crucial,

- Air quality sensors alert people when the air becomes toxic,

- Geo- or persona-enabled text alerts filled with relevant information or instructions will help individuals keep track of the concerned person. This can help older people and young children stay safe,

- Travel companies can track maintenance using sensors to prevent future accidents.

However, thanks to smart appliances — the most important place for an individual where IoT is making the most difference is homes. These IoT security services devices that use artificial intelligence to sense and machine learning to function need constant innovation to stay ahead of the attackers, deflect danger, and keep users safe. The bottom line is even though no technology or smart devices of IoT can always stop bad things from happening. Still, solid acting sensors, can definitely help prevent dangerous instances from escalating.

7.5 Sensor enabling remote health care

A good health care system is the need of every country. It does not matter whether the country is developed, developing, or underdeveloped, when a person needs urgent medical attention, you need the best infrastructure, the best doctors, and diagnostic equipment to timely diagnose a person. However, all these things take time and require a lot of money. It takes years and years of planning, and even then, this approach does not cater to a patient once they leave the vicinity of a hospital. Therefore, in many countries, governed and health care officials have realized the smartest move is to take help from technology. And this is where our sensors come to play.

Interestingly, with the current technological advancements in the technology of the Internet of Things (IoT), healthcare systems are now shifting to innovative remote home healthcare support systems, also known as (ShHeS), even though it is not widely used. Still, the works of this model are being proposed to monitor a patients' health status and receive prescriptions all with just one touch. Therefore, a health monitoring system can play a crucial role in reducing physical contact, the need to go to a hospital, get physical consultation.

A sound health care system is the need of every country. It does not matter whether the country is developed, developing, or underdeveloped when a person is in need of medical and attention; you need the best infrastructure, the best doctors, and diagnostic equipment to timely diagnose a person. However, all these things take time and require a lot of money. It takes years and years of planning, and even then, this approach does not cater to a patient once they leave the vicinity of a hospital. This is why many countries, governed and health care officials have realized the most brilliant move is to take help from technology. And this is where our sensors come to play. In healthcare applications, the sensors, either wearable or embedded, are used to assemble physiological information of the human body such as temperature, pressure rate, electrocardiograph (ECG), electroencephalograph (EEG), and so on if the patient has the sensor attached device on him then it becomes too easy to get health-related data from the patient's body. Additionally, environmental data along with a patient's history can also be recorded. For example, if the weather outside is filled with humidity or rain, these sensing devices will be able to gather this data so that it can help in making meaningful and precise inferences on the health conditions of the patients. Data storage and accessibility also play an essential role in the IoT system. A large amount of data is acquired/recorded from various sources (sensors, mobile phones, e-mail, software, and applications). The data from the sensing described above devices are made available to doctors, caregivers, and authorized parties. Sharing these data with the healthcare providers through the cloud/server allows quick diagnosis of the patients and medical intervention if necessary.

The cooperation between the users, patients, and communication module is maintained for effective and secure transmission. Most IoT systems use a user interface that acts as a dashboard for medical caregivers and performs user control, data visualization, and apprehension. A substantial amount of research has been excavated in today's literature that has published the progress of the IoT system in healthcare, especially when it comes to monitoring, control, security, and privacy. Sensors are the most prominent driver of IoT technology. They serve several purposes, from measuring people, sensing their activities, or measuring the weather of the environment or other inanimate objects. However, ever since IoT has come into the limelight, they're most valuable for "saving lives" by helping identify symptoms of underlying medical conditions, e.g., Parkinson's disease (PD)[19] . Those types of sensors that detect Parkinson's have taken various configuration factors from environmental sensors. Since environmental sensors use printed conductive plastics that can accurately see the concentration of carbon dioxide (CO_2) in the air to bright clothes that integrate tri-axial accelerometers directly into garments, the key to these developments is the increasing technological advancements in microelectromechanical systems.

---

[19] Olutosin Taiwo, Absalom E. Ezugwu, Smart healthcare support for remote patient monitoring during covid-19 quarantine, Informatics in Medicine Unlocked, Volume 20, 2020.

Initial prototyping tools: MEMS sensors and bench testing

MEMS uses microfabrication techniques to integrate circuits and microscopic mechanical components into silicon microchips. In doing so, it is possible to create micro-scale sensors with a range of sensing capabilities. While some research focuses on using MEMS sensors for specific healthcare applications, researchers exploit these technologies to create accessible sensor fusion health monitoring systems.

7.6 Putting it all together

The fact that sensors, ISNS, and IoT devices are bringing a revolution in the world is no secret. Remote devices that can enable these advances and help everything from public safety, security to health care needs is not just a growing trend but also the need of the time around the world. As more and more of these devices become connected to the internet, the ability to manage them remotely will become increasingly important. Choosing a reputable remote access solution is essential to ensure secure and uninterrupted access to these devices and the valuable data that they deliver.

- From home to our mobile devices, to factories, buildings, environments, industrial complexes, automobiles, roads, these devices make us more knowledgeable about our needs, surroundings and can help us make data-driven decisions at our will.

- Sensors improve direct and hyper-local marketing, enhance the quality of their services, and add new features by observing customer responses using IoT devices.

- IoT security services devices that use artificial intelligence to sense and machine learning to function need constant innovation to stay ahead of the attackers, deflect danger, and keep users safe.

Chapter 8: Final Words

"Lots of companies don't succeed over time. What do they fundamentally do wrong? They usually miss the future."

Larry Page

"Any product that needs a manual to work is broken."

Elon Musk

The fact remains that intelligent sensor nodes(ISNs) play major roles in all aspects of our lives and will continue to do so as we adopt smarter technologies. Several generations of ISNs have been introduced in the market to date, and the pursuit of making them smaller, cheaper, enhanced functionality, and more efficient are ongoing. The benefits of their presence and the unintended consequences were covered in the previous chapters. Here we summarize what we know at present and what technological, business, and governmental support are needed to scale our connected world safely and securely.

The next generation of technologies will come from low-power wireless, integration, packaging, AI, and distributed analytics in order to scale to trillions of sensors[20].

8.1 What we know now

Worldwide population growth will put us at around 9 billion people by 2025, further driving the demand for goods. Products are becoming intelligent and connected to the Internet. Product-generated data, which is in massive amounts, is collected on databases, then analyzed, shared, and socialized. The number of Internet-connected devices is forecasted to be 500 billion by the

---

[20] Charlie Wilson, Benefits and risks of smart home technologies," EnergyPolicy, vol. 103, pp. 72–83, April 2017.

year 2025 worldwide. This is about fifty-five times the human population in the same year representing another exponential growth.

In the past decade, the usage of sensors in consumer electronics (e.g., mobile phones, tablets, laptops, cameras, security systems), large industrial plants (e.g., factories, agriculture, and hospitals), automobiles (e.g., passenger cars, trucks transitioning to autonomous vehicles), and medical devices (e.g., wearables, mobile devices), have increased significantly. As a consequence of using sensors, besides the substantial impact on all aspects of our lives, the world population is generating more data than ever before. We will exceed 170 Zettabytes ($1ZB = 10^{21}$ bytes) of data by 2025. A Zettabyte is equal to the amount of data that can be stored on 250 billion DVDs. In turn, this rate of data generation is increasing demands for computation, storage, software, analytics, and sensing even further.

Sensors have enabled us to collect meaningful data from our environment, our goods, and ourselves. This data is converted to information to help us make better-informed, data-driven decisions. The sensors' data is also combined (sensor data fusion) with other data sources to generate the necessary information required for the analytics and context awareness. When we look around us, our mobile phones, cars, factories, stores, roads, farms, healthcare, and buildings are helping us to proactively know and do something about our lives, and this is only the tip of the iceberg.

Embedded intelligent sensors will continue to occupy more of our environments and our lives with more uniform, standardized interfaces for ease of use and affordability so that the playing field is leveled.

As catastrophic events such as pandemics, global warming, and energy shortages continue to persist, smart sensors find their ways in these applications to sense, detect, and inform us about these events and our safety.

8.2. Technological Advances needed for the next generations

If we simplify things, we can say that there have been three fundamental industrial revolutions. The first was the steam engine, the second was the assembly line, and the third was the computers.

And, with computers, the term 'technology' was born. It was just a matter of time when these computers went through a change as well, and the greatest global impact on the current generation would start. Today there is a fourth revolution that is taking place, and it is called Industry 4.0. The simplest definition of Industry 4.0 would be making all manufacturing equipment safe, digital, and intelligent. Technology provides some great opportunities for global development and a promising future. It's so powerful that it has given us the opportunity to satisfy many of our short-term needs and wants instantly, the proof of which has been work-from-home during the COVID-19 pandemic. However, for the next generation of professionals to succeed, they must stay up to date with the latest tech, innovations, and tools. Developing technologies, such as self-driving cars, industrial robots, artificial intelligence, and decision-making gadgets are advancing at a rapid pace. While "blockchain," "artificial intelligence," and "drones" may be the current buzzwords surrounding tech in global development, geographical information systems, or GIS, and big data are actually the top technologies that I believe the next generation of development professionals should learn how to utilize.

For us, IoT has shown a drastic performance to change the way we live. The technology of sensors has undergone a continuous change. With time new advancements have made life more efficient, comfortable, and reliable. IoT products are, in a sense, self-aware. They can share information about their health, location, and usage levels. This, in turn, helps companies improve product quality and customer service.

For example, IoT devices can track shipments in real-time, while AI technology can route trucks based on current road conditions. Automated document processing can speed goods through customs. And today, with the likes of Tesla, some companies are developing fleets of self-driving trucks. Several ports worldwide have introduced automated cranes and guided vehicles that can unload, stack, and reload containers faster and with fewer errors. We will see more advanced development of optical MEMS devices and systems. In the very long term, we may possibly develop nanoelectromechanical systems (NEMS) or Optical NEMS[21].

Therefore, IoT has a lot to help in multiple aspects of life and technology. We may assume that IoT has a lot of scope, both in terms of technology enhancement and performance improvement, in facilitating humankind.

Given all the progress that has been made, we are still at the infancy stage in many of the critical issues that we are struggling with worldwide. For example, the COVID-19 pandemic has increased efforts to develop sensor technology to manage disease detection. As a result of efforts, the sensor technology definition is being brooded to include physical, cellular, and molecular platforms that produce signals to identify specific events associated with SARS-CoV-2 and/or its interaction with the host.

The main applications of sensor technology in COVID-19 have been to detect a fever using touchless infrared thermometers and the presence of viral RNA using polymerase chain reaction (PCR). However, a substantial proportion of individuals with COVID-19 never develop a fever. PCR tests have been developed to detect SARS-CoV-2 in nasopharyngeal samples, but to date, they have been expensive, resource-intensive, cumbersome, and relatively slow. Moreover, positive PCR tests do not imply a person is still infectious and thus have not provided information about transmissibility or virulence, hampering the development of more effective action plans that are scalable. We are at the early state of sensing technologies for SAR-CoV-2. Recently FDA has approved Antigen testing that can be performed using an at-home kit. Although this is a massive improvement, we are still lacking a

---

[21] Stojkoska, B. L. R., &Trivodaliev, K. V A review of Internet of Things for smart home: Challenges and solutions. Journal of Cleaner Production, 140, 1454-1464. Inertial Sensor Technology Trends Neil Barbour and George Schmidt, Senior Member, IEEE, 2017.

real-time robust test with high certainty. This is where the government and large organizations need to focus since there are both economic and societal long-term impacts.

Other examples of areas that need special attention include water/air pollution detection, global warming, earthquake/hurricane/tsunami warning systems. Unless there is massive governmental and private enterprise support, the progress will not be meaningful.

8.3 Business changes needed for the next generations

The creation of next-generation products and their derivatives seems to be the routine work of technology-based companies; it's the setup their entire business model is built on. The ability to accommodate and readjust to fast-paced business change has become critical for competitiveness and growth. There are several ways companies can occasionally increase their performance, but none is as necessary over the long term as the development of an ability to sustain innovation. There have been extensive research efforts to understand the drivers behind the capacity to innovate, especially in innovation management and design theories continuously.

Next-generation business ideas will heavily be influenced by the current GEN Z, who, as people call them 'the generation that was born to be digital,' are the product of the growing interconnectedness they are immersed in. Therefore changes in the internal and external setup of businesses will heavily be focused on the transition to a next-generation operating model. As GEN-Z relies heavily on Smartphones, Social media, Virtual reality, Artificial intelligence, and the cloud so it makes sense that the next generational business ideas will follow along the path of those two concepts.

Businesses with technological products are developing round the clock to serve the future needs of customers. However, newer technology does not always warrant immediate success. Instead, it is a risky business move because companies are at high risk when they work with new, unproven technologies or architectures, but since the industry is highly competitive, the only way to stay ahead is with innovation. That's one business setup future generations must get ready for.

The present and future currency is information that is derived from comprehensive and massive data sets. Tech companies, along with other industries, use data to generate deeper insights, discover opportunities, and make better, more timely decisions. Hence, the companies that efficiently do this with new data will lead the next generation. However, it won't be that easy. With Companies like Apple changing their data policies, people know giving up their data is not a smart thing. Instead of flourishing from the data they had collected, companies are now taking pride in how secure they are by keeping their user's data private. So, maybe the future of technology and data is not as "public" as it once seemed.

8.4 Technology and Business interplay

A robust cloud/enterprise/swarm framework and technical advances are not enough. To stand out and catch up with our own needs, we have to realize that the only solution is an interplay of these entities. This is proven by history. If we look back, we can see that the interplay between business and information systems engineering (BISE)[22] and its primary reference disciplines, business administration, and informatics, has changed several times during the last decades. Important examples of this "interplay" include:

- Data management and modeling

- Internet-based information and transaction systems

- The emergence of computational business sciences

---

[22] Smajlović, Selma & Umihanic, Bahrija & Turuljathe, L & Smajlović, Selma & Umihanić, Bahrija & Turulja, Lejla. THE INTERPLAY OF TECHNOLOGICAL INNOVATION AND BUSINESS MODEL INNOVATION TOWARD COMPANY PERFORMANCE. 24. 63-79. 10.30924/mjcmi.24.2.5. (2019).

The increasing integration of the virtual and real-world in the "Internet of Things" offers new scientific and organizational challenges for the cooperation between the BISE and informatics, which can only be formulated and solved jointly.

Integrating the virtual and real-world in the "Internet of Things" offers new scientific and organizational challenges for this cooperation. In conclusion, this book accentuates the need to establish a proactive innovation culture within an organization aligned with the growing demand in the technology-based industry. It further suggests a need to implement organizational changes relevant to radical innovation of technologies and advances that are happening as we speak. Secondly, this interplay between technology and business models is essential for future research to gain more clarity on understanding the different management methods that will help us deliver better technological advances outcomes within a digitalized business environment.

Lastly, at the time of this writing, the Covid pandemic is still lingering and will stay with us for a few more years. What is observed is that solutions to such problems come mainly by partnership and collaboration between researchers and the broader community from the outset; the collaborations need to include shared priorities of patients, the community, health professionals, scientists, engineers, and policymakers. These collaborations will enable and prioritize research in sensor technology to address COVID-19 and can facilitate clinicians and diagnosticians to evaluate, diagnose and treat patients, manage remote behavioral monitoring and emergency response without any physical interaction through the application of IoT-based platforms. This technological play can significantly help to monitor and manage the pandemic both at personal and organizational levels.

Appendices

Sensor / Technology/Business Glossary
List of references

[1]. R. Spence, "Top Driverless Trucks in the Mining Industry Today Plus Future Concepts," Mining Global, 07 July 2014 2014.

[2]. K. Diss, "Driverless trucks move all iron ore at Rio Tinto's Pilbara mines, in world first," 2015.

[3]. J. A. Rodger, "Toward reducing failure risk in an integrated vehicle health maintenance system: A fuzzy multi-sensor data fusion Kalman filter approach for IVHMS," Exp. Syst. App. 39, 9821-2012.

[4]. J. A. Rodger, "Application of a Fuzzy Feasibility Bayesian Probabilistic Estimation of supply chain backorder aging, unfilled backorders, and customer wait time using stochastic simulation with Markov blankets," Exp. Syst. App., vol. 41, pp. 7005-7022, 11/15/ 2014.

[5]. NASA, Real-Time Tracking System Uses Ultra-Wideband RF Signals," 2012.

[6]. K. Hlophe and F. du Plessis, "Implementation of an autonomous underground localization system," RobMech, 87-92, 2013

[7]. "Scanning Range Sensor Performance for Mining Automation Applications," CRC Mining Latest News2014.

[8]. E. Ruth, "Mining truck spotting under a shovel," Google Patents, 2014.

[9]. D. Tadic, "CRCMining cave tracker system - pioneering flow monitoring for caving," AusIMM Bulletin, vol. 6, pp. 80-81, 2014.

[10]. Apple iBeacons explained–smart home occupancy sensing solved? 2013.

[11]. N. J. Lavigne, J. A. Marshall, U. Artan, and Ieee, " Towards underground mine drift mapping with RFID," 23rd Canadian Conference on Electrical and Computer Engineering, 2010.

[12]. Avoiding accidents with mining vehicles [Online]. Available: http://www.flir.com/cs/emea/en/view/?id=51907

[13]. S. Bagavathiappan, B. B. Lahiri, T. Saravanan, J. Philip, and T. Jayakumar, "Infrared thermography for condition monitoring – A review," Infrared Physics & Technology, vol. 60, pp. 35-55, 2013.

[14]. "FLIR cameras enable timely detection and localisation of selfcombusting coals," Australian Mining, 26 June, 2014.

[15]. "Advances in detectors: Hot IR sensors improve IR camera size, weight, and power," Laser Focus World, 17 January 2014.

[16]. D. Szondy, "DARPA developing personal LWIR cameras to give soldiers heat vision," 2013.

[17]. M. Magno, D. Boyle, D. Brunelli, E. Popovici, and L. Benini, "Ensuring Survivability of Resource-Intensive Sensor Networks Through UltraLow Power Overlays," IEEE Trans. Ind. Inf., 10, pp. 946-956, 2014.

[18]. L. Li, "Time-of-Flight Camera – An Introduction," Texas Instruments Technical White Paper, vol. SLOA190B, 2014.

[19]. "3D Time of Flight Imaging Solutions," Texas Instruments 2014.

[20]. J. Han, L. Shao, D. Xu, and J. Shotton, "Enhanced computer vision with microsoft kinect sensor: A review," 2013.

[21]. M. Hiramoto, Y. Ishii, and Y. Monobe, "Light field image capture device and image sensor," ed: Google Patents, 2014.

[22]. "XRF Analyzer and XRD Analyzer Solutions: Geochemistry and Mining," 2014.

[23]. "Miniature Mossbauer spectrometer. Applications to mining," 2014.

[24]. R. V. Morris, G. Klingelhöfer, B. Bernhardt, C. Schröder, D. S. Rodionov, P. A. de Souza, et al., "Mineralogy at Gusev Crater from the Mössbauer Spectrometer on the Spirit Rover," Science, 305, 833, 2004.

[25]. A. Malcolm, S. Wright, R. R. A. Syms, R. W. Moseley, S. O'Prey, N. Dash, et al., "A miniature mass spectrometer for liquid chromatography applications," Rap. Comm. Mass Spectr., 25, pp. 3281-3288, 2011.

[26]. M. Dalm, M. W. N. Buxton, F. J. A. van Ruitenbeek, and J. H. L. Voncken, "Application of near-infrared spectroscopy to sensor based sorting of a porphyry copper ore," Minerals Eng., 58, pp. 7-16, Apr 2014.

[27]. J. Lessard, J. d. Bakker, and L. McHugh. Development of ore sorting and its impact on mineral processing economics. Minerals Eng. 88, 2014

[28]. M. Dransfield, "Airborne Gravity Gradiometry in the Search for Mineral Deposits," Advances in Airborne Geophysics, vol. 20, 2007.

[29]. J. Anstie, T. Aravanis, M. Haederle, A. Mann, S. McIntosh, R. Smith, et al., "VK-1 - a new generation airborne gravity gradiometer," ASEG Extended Abstracts, vol. 2009, pp. 1-5, 2009.

[30]. D. DiFrancesco, T. Meyer, A. Christensen, and D. FitzGerald, "Gravity gradiometry - today and tomorrow," 11th SAGA Biennial Technical Meeting and Exhibition, Swaziland, 2009.

[31]. "4D Seismic Comes of Age," Offshore Technology2008.

[32]. (2014). What is Microseismic Monitoring? Available: https://www.esgsolutions.com/english/view.asp?x=852

[33]. W. T. Pike, I. M. Standley, and S. Calcutt, "A silicon micro-seismometer for Mars," Transducers & Eurosensors, 622, 2013

[34]. "Health Monitoring in Mining Code of Practice," Safe Work Aust., 2011.

[35]. S. Patel, H. Park, P. Bonato, L. Chan, and M. Rodgers, "A review of wearable sensors and systems with application in rehabilitation," J. Neuroeng. Rehab., vol. 9, p. 21, 2012.

[36]. A. Pantelopoulos and N. G. Bourbakis, "A survey on wearable sensorbased systems for health monitoring and prognosis," Trans. Sys. Man Cyber Part C, vol. 40, pp. 1-12, 2010.

[37]. "HASARD - An Electromagnetic-Based Proximity Warning System " Office of Mine Safety and Health Research, 2014.

[38]. D. Chirdon, "Proximity Detection / Collision Warning," Mine Safety and Health Administration, 2014.

[39]. A. Costanzo, M. Dionigi, D. Masotti, M. Mongiardo, G. Monti, L. Tarricone, et al.,"Electromagnetic Energy Harvesting and Wireless

[40]. Budman, M., Khan, A., and Cotteleer, M., "Forces of change: Industry 4.0," Deloitte Insights, 1 –20 (2017). Google Scholar

[41]. Coal-Fired Electricity Generation in the United States and Future Outlook," MJB&A Issue Brief, 1 –14 (2017). Google Scholar

[42]. "Distributed Energy Resources: Connection Modeling and Reliability Considerations," North American Electric Reliability Corporation, Distributed Resources Task Force Report, (2017). Google Scholar

[43]. "DOE Laboratories Sign Memorandum of Understanding for Innovative Coal Research," (2018). Google Scholar

[44]. EEI Finance Department, Company Reports, S&P Global Market Intelligence, (2016). Google Scholar

[45]. "Future of Energy: Digital for Coal-Fired Plants," GE Power Digital Solutions, (2017). Google Scholar

[46]. Grant, C., McCue, J., and Young, R., "The Power is On: How IoT Technology is Driving Energy Innovation," Deloitte Center for Energy Solutions, 1 –24 Deloitte University Press,2016). Google Scholar

[47]. Grol, E., Tarka, T., Myles, P., Bartone, L, Simpson, J., and Rossi, G., "Impact of Load Following on the Economics of Existing Coal-Fired Power Plant Operations," DOE/NETL – 2015/1718, (2015). Google Scholar

[48]. Hiskey, T., "Preparing for Manufacturing's Future with Industry 4.0," Industry Week, (2018) Google Scholar

[49]. "Investigation of Smart Parts with Embedded Sensors for Energy System Applications," National Energy Technology Laboratory, Project Description, (2018) Google Scholar

[50]. Lockwood, T., "Advanced sensor and smart controls for coal-fired power plant," (21), IEA Clean Coal Center, No.2015). Google Scholar

[51]. Louis, M., Seymore, T., and Joyce, J., "3D Opportunity in the Department of Defense: Additive Manufacturing Fires Up," Deloitte University Press, (2014). Google Scholar

[52]. Miller, D. and Henly, C., "Blockchain is Reimagining the Rules of the Game in the Energy Sector," Rocky Mountain Institute, (2017) Google Scholar

[53]. Starks, T., "U.S. says Russian hackers targeted American energy grid," Politico, (2018). Google Scholar

[54]. Stern, D., "The Role of Energy in Economic Growth," Crawford School of Climate Economics and Policy, (2011). Google Scholar

[55]. U.S. EIA, "Most coal plants in the United States were built before 1990.," (2017). Google Scholar

[56]. Essential Principles of Image Sensors 1st Edition by Takao Kuroda, ISBN-13: 978-1482220056;

[57]. Bhalla, N., Pan, Y., Yang, Z. & Payam, A. F. *ACS Nano* **14**, 7783–7807 (2020).

[58]. Budd, J. et al. *Nat. Med.* **26**, 1183–1192 (2020).

[59]. Koff, W. C. & Williams, M. A. N. *Engl. J. Med.* **383**, 804–805 (2020).

[60]. Barbour, Neil & Schmidt, George. Inertial Sensor Technology Trends. Sensors Journal, IEEE. 332 – 339 (2002).

[61]. A S Abdul-Qawy et al. Int. Journal of Engineering Research and Applications www.ijera.com ISSN: **2248-9622,** Vol. 5, Issue 12, (Part - 2) pp.71-82 December (2015).

[62]. Kaur, N.; Sood, S.K. An energy-efficient architecture for the Internet of Things (IoT). IEEE Syst. J., **11**, 796–805, (2015).

Bibliography

[1]. D. K. Singh, R. Desai, N. Walde, P. B. Karandikar, "Nano warehouse: A New Concept for Grain Storage in India," 2014 International Conference on Green Computing Communication and Electrical Engineering, 2014.

[2]. Z. Pang, Q. Chen, W. Han, and L. Zheng, "Value-centric design of the internet-of-things solution for food supply chain: Value creation, sensor portfolio and information fusion," Information Systems Frontiers, vol. 17, pp. 289-319, 2015.

[3]. Smajlović, Selma & Umihanic, Bahrija & Turuljathe, L & Smajlović, Selma & Umihanić, Bahrija & Turulja, Lejla. The Interplay of Technological Innovation and Business Model Innovation Toward Company Performance. 24. 63-79. 10.30924/mjcmi.24.2.5. (2019).

[4]. Bhandari, S., Gangola, P., & Verma, S. IoT BASED FOOD MONITORING SYSTEM IN WAREHOUSES. International Research Journal of Engineering and Technology (IRJET). https://www.irjet.net/archives/V7/i4/IRJET-V7I41124.pdf. (2020).

[5]. Warehouses and Cold Storescomprehensive Monitoring of Temperature and Humidity. Monitoring Temperature and Humidity (2021).

[6]. Laura Stevens, Amazon Delays Opening of Cashierless Store to Work Out Kinks, Wall Street Journal, March 27, 2017.

[7]. Autonomous Truck Market Size to Hit 2,013.34 Million by 2027; Rising Installation of Advanced Radars, Sensors & Cameras will Boost Growth, Says Fortune Business Insights (2019).

[8]. Russo, G., Marsigalia, B., Evangelista, F., Palmaccio, M., &Maggioni, M. Exploring regulations and scope of the Internet of Things in contemporary companies: a first literature analysis. Journal of Innovation and Entrepreneurship, (2015).

[9]. Viktor Gubochkin, How To Use IoT For Smart Parking Solution Development, April 14, 2020.

[10]. Kaur, N.; Sood, S.K. An energy-efficient architecture for the Internet of Things (IoT) 11, 796–805.. IEEE Syst, 2015.

[11]. Pasika, S., & Gandla, S. T. Smart water quality monitoring system with cost-effective using IoT. Heliyon, 6(7), (2020).

[12]. Sow, D. et al. "Mining of Sensor Data in Healthcare: A Survey." Healthcare Data Analytics (2015).

[13]. Ayyakkannu Purushothaman, Comparison of Manual Versus Automated Data Collection Method for Haematological Parameters, February 19, 2019.

[14]. Ayyakkannu Purushothaman, Comparison of Manual Versus Automated Data Collection Method for Haematological Parameters, February 19, 2019.

[15]. Olutosin Taiwo, Absalom E. Ezugwu, Smart healthcare support for remote patient monitoring during covid-19 quarantine,Informatics in Medicine Unlocked,Volume 20,2020.

[16]. Olutosin Taiwo, Absalom E. Ezugwu, Smart healthcare support for remote patient monitoring during covid-19 quarantine, Informatics in Medicine Unlocked,Volume 20, 2020.

[17]. Charlie Wilson, Benefits and risks of smart home technologies," EnergyPolicy, vol. 103, pp. 72–83, April 2017.

[18]. Stojkoska, B. L. R., &Trivodaliev, K. V A review of Internet of Things for smart home: Challenges and solutions. Journal of Cleaner Production, 140, 1454-1464. Inertial Sensor Technology Trends Neil Barbour and George Schmidt, Senior Member, IEEE, 2017.

[19]. Smajlović, Selma & Umihanic, Bahrija & Turuljathe, L & Smajlović, Selma & Umihanić, Bahrija & Turulja, Lejla. THE INTERPLAY OF TECHNOLOGICAL INNOVATION AND BUSINESS MODEL INNOVATION TOWARD COMPANY PERFORMANCE. 24. 63-79. 10.30924/mjcmi.24.2.5. (2019).

[20]. Bruno Aziza ,What Customers Teach Us About The Future Of Data, Jun 3, 2021.

Page Left Blank Intentionally

ARTIFICIAL SENSORS

Made in the USA
Las Vegas, NV
08 October 2021